Selenium in Natural Produ Synthesis

Selenium in Natural Products Synthesis

K. C. Nicolaou
N. A. Petasis

Department of Chemistry
University of Pennsylvania
Philadelphia, Pennsylvania

Foreword by
K. B. Sharpless

Department of Chemistry
Massachusetts Institute of Technology
Cambridge, Massachusetts

CIS *CIS, INC.* *Philadelphia* *1984*

PUBLISHED BY:
CIS, Inc.
P. O. Box 7741, Philadelphia, PA 19101

ISBN 0-914891-00-6
Library of Congress Catalog Card No. 83-72883

PRINTED IN THE UNITED STATES OF AMERICA

Foreword

Three decades ago, organic chemists greeted with enthusiasm Georg Wittig's publication describing a mild and selective method for olefin synthesis, for the method's potential utility was immediately apparent.

As useful as the Wittig reaction was—and remains today—it had a much larger impact as well. Organic chemists whose thoughts had rarely slipped below the first row of the periodic table began to ponder phosphorus's "exotic" neighbors, elements which might give rise to other new and useful reagents. The legacy of this prospecting in the exotic outskirts of the periodic table continues to grow, and, indeed, the ancestry of most modern reagents and much of organometallic chemistry can be traced to the Tübingen laboratory of Georg Wittig.

Selenium was a little *too* exotic. Many of the organosulfur reagents in use today were already widely used before investigators turned to selenium. Its known toxicity called for caution; its malodorous reputation made it doubly unattractive to any chemist who wanted a social life outside the lab.

Despite these drawbacks, the discovery in the early 1970's that selenoxides eliminate to form olefins under exceedingly mild conditions quickly gave rise to a large organoselenium methodology. Like the Wittig reaction, organoselenium-based methods offer mild and selective routes to olefins as their primary virtue, and, at this writing, over a thousand publications utilizing new organoselenium reagents have appeared in print. Furthermore, selenium-based technology today extends its usefulness to a large variety of other synthetic operations.

Judging the real utility of these reagents, however, is only possible *via* an evaluation of how they perform in especially restrictive circumstances, like those encountered in the synthesis of complex natural products, and this volume does an excellent job of fulfilling that criterion. The authors' adept selection of cases and demonstrations reflects the position of Professor Nicolaou's group at the forefront of both new organoselenium methodology and complex natural products synthesis. Their book, the first devoted to selenium reagents in organic synthesis, will be much used.

Cambridge, Massachusetts K. B. Sharpless
March, 1984

Preface

Organic synthesis has evolved into a fascinating combination of art, science and technology. At its best, it utilizes the creativity and elegance of art, the precise execution and discipline of science and the ingenuity and pragmatism of technology. Furthermore, it serves as a valuable aid towards our understanding of nature, since nature has been practicing organic synthesis for a very long time. The number and variety of organic compounds synthesized by nature is impressive. The structures of most of these compounds are often intellectually appealing and challenging and it is, therefore, not surprising that natural products have always been among the most popular targets of synthetic chemists.

In designing the total synthesis of a complex molecule, it is usually necessary to utilize highly selective transformation sequences. Developing efficient transformations is by itself a very worthwhile task. Indeed, in the last decade or so, we have witnessed an outburst of new discoveries in synthetic methodology, many of them characterized by remarkable selectivity and mildness. The exploration of the chemistry of various "uncommon" elements has produced many of these methods.

In this book we attempt to demonstrate the varsatility and overall synthetic utility of the various reagents containing selenium. This element, due to its mixed metallic and non-metallic nature and its unique reactivity towards oxidation and reduction, has provided the synthetic chemist with a large number of valuable synthetic tools. In addition to recent examples of the well-known selenium dioxide oxidations, the book covers extensively the numerous synthetic applications of the newly developed organoselenium reagents. The various transformations are classified according to reaction type. After a brief description, each method is illustrated with examples from natural products synthesis.

Although the main focus of the book is selenium-based methodology, significant attention is devoted to the overall synthetic schemes. Thus, the particular step employing selenium is placed in context and in proper perspective so that the reader can appreciate the strategy of the synthesis as well. While in some cases only a small segment of the total synthesis is given, in many instances a larger segment is presented, in order to give the reader the opportunity to review some of the outstanding achievements in natural products synthesis.

The references include papers published up to the end of 1983. It is inevitable, however, that in such a widespread field, some references have been omitted inadvertently, and for that we apologize. Nevertheless, we believe that enough evidence is provided to encourage further applications and to stimulate new ideas.

We hope that this book will serve the practicing organic chemist as a useful source of information for synthetic applications of organoselenium chemistry. In addition, we believe that students and teachers of organic synthesis will find it helpful in advancing their understanding of the strategies used in the total synthesis of natural products.

We express our thanks and sincere appreciation to Professor K. B. Sharpless, one of the pioneers of the chemistry described herein, for his kindness in writing the foreword and providing valuable comments and recommendations. We are also indebted to Professor M. M. Joullié for reading the manuscript and making many useful comments, and to Dr. M. J. Mitchell for helpful suggestions on style and format. Finally, we extend our deep appreciation to our many coworkers, past and present, for their enthusiasm and dedication, which led to the success of various research projects, some of which are described in this book.

Philadelphia, Pennsylvania K. C. Nicolaou
March, 1984 N. A. Petasis

ABBREVIATIONS

Ac	Acetyl
acac	Acetylacetonyl
AIBN	Azobisisobutyronitrile
AcOH	Acetic acid
AcOOH	Peracetic acid
Bz	Benzoyl
12-C-4	12-Crown-4 ether
18-C-6	18-Crown-6 ether
cat	Catalyst
CSA	Camphorsulphonic acid
DABCO	1,4-Diazabicyclo[2.2.2]octane
DBU	1,5-Diazabicyclo[5.4.0]undec-5-ene
DDQ	2,3-Dichloro-5,6-dicyano-1,4-benzoquinone
DIBAL	Diisobutylaluminum hydride
DMAN	Dimethylaminonaphthalene
DMAP	4-Dimethylaminopyridine
DME	1,2-Dimethoxyethane
DMF	*N,N*-Dimethylformamide
DMSO	Dimethylsulphoxide
eq	Equivalent
HETE	Hydroxyeicosatetraenoic acid
HMPA	Hexamethylphosphoramide (hexamethylphosphoric triamide)
KDA	Potassium diisopropylamide
LDA	Lithium diisopropylamide
LDBA	Lithium diisobutylamide
LICA	Lithium isopropylcyclohexylamide
LTMP	Lithium 2,2,6,6-tetramethylpiperidine
mCPBA	*m*-Chloroperbenzoic acid
MICA	Magnesium isopropylcyclohexylamine
Ms	Mesyl (methylsulfonyl)
NBS	*N*-Bromosuccinimide
NCS	*N*-Chlorosuccinimide
N-PSP	*N*-Phenylselenosuccinimide
PCC	Pyridinium chlorochromate
Ph	Phenyl
n-Pr	*n*-Propyl
pyr	Pyridine
Tf	Triflic (trifluoromethanosulfonyl)
THF	Tetrahydrofuran
TMEDA	*N,N,N,N*-Tetramethylenediamine
Ts	Tosyl (*p*-toluenesulfonyl)
TsOH	Tosic acid (*p*-toluenesulfonic acid)
Δ	Heat

Contents

CHAPTER 1

Introduction: Selenium Compounds in the Synthesis of Natural Products

The development in recent years of a large number of selenium-based synthetic methods has made significant contributions to synthetic organic chemistry. Of special value is the application of selenium methodology to the synthesis of natural products, which often requires highly selective and very efficient transformations. Indeed, as will be demonstrated in what follows, selenium reagents have been used on numerous occasions to perform key synthetic operations.

1.1. The challenge of natural products synthesis

Naturally occurring compounds have long fascinated the organic chemist. In recent years the isolation and structural elucidation of natural products have been enormously facilitated by the development of chromatography, spectroscopy and X-ray crystallography, with the result that an ever growing number of new structures are finding their way into the chemical literature. The total synthesis of natural products was originally aimed at confirming their molecular structure, but more recently it has been the biological profile combined with structural novelty that have been the impetus for developing a total synthesis. Whatever the motive, the endeavor itself can provide unique opportunities for creative research, which often leads to the discovery and application of new chemical reactions and novel reagents.

Indeed, in an effort to find mild and selective ways to perform problematical operations, synthetic chemists have recently been exploring the organic chemistry of some chemical elements that were previously relatively unknown. As a result, a number of powerful synthetic methods have been evolved and many new reagents have been discovered. Compounds of elements such as boron, aluminum, silicon, tin, phosphorus, sulfur, selenium, palladium, copper, zirconium, cobalt, titanium and others, are now in the arsenal of the organic chemist. The selenium-based reagents in particular have provided useful methods for the attachment of oxygen, for the introduction of olefinic bonds and for the cyclization of unsaturated systems, as well as for carrying out various other transformations. In many instances, the use of a particular selenium-containing reagent was found to be indispensable to the success of a multistep synthesis.

1.2. Some unique properties of selenium compounds

Selenium and its compounds are by no means new in organic chemistry. Elemental selenium was used for a long time for the dehydrogenation of polycyclic compounds, as a means of structure elucidation. In addition, selenium dioxide has been widely recognized as a powerful reagent for effecting selective oxygenations and dehydrogenations. The chemistry and applications of selenium dioxide have been thoroughly reviewed.[1]

Organoselenium compounds have also been known for over a century and their chemistry has been extensively studied and reviewed.[2] In the past decade, however, organoselenium chemistry has acquired a new dimension.[3-9] The pioneering work of Sharpless[4] and Reich[5,6] has uncovered the unique ability of the phenylseleno group to act as a handle for the introduction of olefinic bonds. It was also found by Barton and coworkers[7] that benzeneseleninic anhydride is a versatile oxidizing agent, complementary to selenium dioxide. Furthermore, it was recognized by Nicolaou,[8] Clive,[10] Ley[11] and others, that certain electrophilic organoselenium reagents can induce ring closure with high regio- and stereoselectivity, leading to various heterocyclic and carbocyclic systems. Another major advance in the field stems from the elegant work of Reich,[5,6] Krief[9] and others on the utility of selenium-stabilized carbanions.

The new wave of selenium reasearch has changed for the better the reputation of selenium reagents, which were formerly thought of as "nasty," dangerous and undesirable. Although most selenium compounds are in fact toxic and malodorous, they can be used safely, like so many other "hazardous" reagents, by adopting the proper techniques of handling and disposal. Selenium reagents have now been widely accepted, particularly for small-scale reactions such as those encountered in natural products synthesis.

Several useful comparisons can be made between selenium and sulfur compounds.[6] In general, the oxidation of selenides to selenoxides proceeds more readily than the oxidation of sulfides to sulfoxides. However, in the case of selenium, the second oxidation level, *i.e.* the formation of selenones, is attained with considerable difficulty, as compared with the oxidation of sulfoxides to sulfones. Another major difference between the two elements lies in the strength of their bonds to carbon. The C–Se bond is weaker, compared to the C–S bond, consequently several reactions involving fission of this bond proceed much easier with the selenium compounds. Examples are: (a) the *syn*-elimination of selenoxides to form olefins, (b) the reductive cleavage of selenides to hydrocarbons and (c) the [2,3]-sigmatropic rearrangement of β,γ-unsaturated selenoxides. These properties often make the methodology of selenium reagents preferable to that of the corresponding sulfur reagents.

1.3. Applications of selenium reagents in natural products synthesis

The most common selenium reagents used in organic synthesis are:

Se	PhSeCN and *o*-$NO_2C_6H_4SeCN$
SeO_2	$PhSeSO_2Ar$
PhSe(O)OSe(O)Ph	PhSeH
PhSe(O)OOH	$(PhSe)_3B$
$Ph_2Se(OCOCF_3)_2$	$PhSeSiMe_3$
PhSeSePh	MeSeSeMe
PhSeCl	N-Phenylselenosuccinimide (N-PSS)
PhSeBr	N-Phenylselenophthalimide (N-PSS)

Many of the above reagents are now commercially available (from Aldrich, Alfa, Fluka, etc.), which should encourage their use. As will become evident in the chapters that follow, they have been used to good advantage in the synthesis of many target molecules, and there is almost no restriction to the kinds of compounds that can be synthesized. The selenium-based methodology has been utilized in numerous syntheses of natural products of all types, including steroids, terpenoids, eicosanoids, alkaloids, carbohydrates and various antibiotics.

REFERENCES

1. (a) Rabjohn, N., "Selenium Dioxide Oxidation" in Dauben, W. G. (ed.), *Organic Reactions*, Wiley, New York (1976), Vol. 24, Chap. 4, p. 261; (b)

Jerussi, R. A., in Thyagarajan, B. S. (ed.), *Selective Organic Transformations*, Wiley, New York (1970), Vol. 1., p. 301; (c) Trachtenberg, E. N., "Selenium Dioxide Oxidation" in Augustine, R. L. (ed.), *Oxidation Techniques and Applications in Organic Synthesis*, Dekker, New York (1969), Vol. 1, Chap. 3, p. 119; (d) Rabjohn, N., "Selenium Dioxide Oxidation" in Adams, R., (ed.), *Organic Reactions*, Wiley, New York (1949), Vol. 5, Chap. 8, p. 331; (e) Waitkins, G. R., and Clark, C. W., *Chem. Rev.*, **36**, 235 (1945).

2. (a) Irgolic, K. J., and Kudchadker, M. V., "Organic Chemistry of Selenium," in Zingaro, R. A., and Cooper, W. C., (eds), *Selenium*, Van Nostrand Reinhold, New York (1974), Chap. 8, p. 408; (b) Klayman, D. L., and Gunther, W. H. H., (eds.), *Organic Selenium Compounds: Their Chemistry and Biology*, Wiley, New York (1973); (c) Gosselck, J., *Angew. Chem. Int. Ed. Engl.*, 2, 660 (1963); (d) Campbell, T. W., Walker, H. G., and Coppinger, G. M., *Chem. Rev.*, **50**, 279 (1952).
3. (a) Clive, D. L. J., *Tetrahedron*, **36**, 2531 (1980); (b) Clive, D. L. J., *Aldrichimica Acta*, **11**, 43 (1978).
4. Sharpless, K. B., Gordon, K. M., Lauer, R. F., Patrick, D. W., Singer, S. P., and Young, M. W., *Chem. Scr.*, **8A**, 9 (1975).
5. Reich, H. J., "Organoselenium Oxidations," in Trahanovsky, W. S. (ed.), *Oxidation in Organic Chemistry, Part C*, Academic Press, New York (1978), Chap. 1, p. 1.
6. Reich, H. J., *Acc. Chem. Res.*, **12**, 22 (1979).
7. Barton, D. H. R., and Ley, S. V., in *Further Perspectives in Organic Chemistry*, Elsevier, New York (1978).
8. Nicolaou, K. C., *Tetrahedron*, **37**, 4079 (1981).
9. Krief, A., *Tetrahedron*, **36**, 2531 (1980).
10. (a) Clive, D. L. J., Chittattu, G., Curtis, N. J., Kiel, W. A., and Wong, C. K., *J. C. S. Chem. Comm.*, 725 (1977); (b) Clive, D. L. J., Russell, C. G., Chittattu, G., and Singh, A., *Tetrahedron*, **36**, 1399 (1980); (c) Clive, D. L. J., Farina, V., Singh, A., Wong, C. K., Kiel, W. A., and Menchen, J. M., *J. Org. Chem.*, **45**, 2120 (1980).
11. (a) Jackson, W. P., Ley, S. V., and Morton, J. A., *J. C. S. Chem. Comm.*, 1028 (1980); (b) Jackson, W. P., Ley, S. V., and Morton, J. A., *Tetrahedron Lett.*, 22, 2601 (1981).

CHAPTER 2

Selenium-mediated Oxygenations

The best-known selenium compound in organic synthesis is selenium dioxide, SeO_2, which has been used traditionally as an oxidizing agent for olefins, ketones and other compounds.[1] Due to its unique abilities, selenium dioxide continues to be a synthetically valuable reagent. Furthermore, some powerful new procedures have made the use of selenium dioxide more practical and more effective. Among the most important of these is a method introduced by Sharpless[34] which utilizes catalytic or stoichiometric amounts of selenium dioxide in the presence of *t*-butyl hydroperoxide (^{t}BuOOH). Complementary to selenium dioxide is benzeneseleninic anhydride, PhSe(O)OSe(O)Ph, a novel oxidizing agent recently exploited by Barton.[54, 65, 67, 69] Also, benzeneperoxyseleninic acid, PhSe(O)OOH, is a novel reagent for the epoxidation of olefins[74] and for Baeyer–Villiger reactions.[80]

The above selenium-containing reagents are among the most important means of introducing oxygen into organic molecules. Oxygenation reactions involving these reagents are the subject of this chapter.

2.1 Allylic oxidation of olefins

One of the unique properties of selenium dioxide is its ability to introduce an oxygen at the allylic position of olefins (1). Depending on the reaction conditions, the major product is either the allylic alcohol (2) (or acetate) or the corresponding α,β-unsaturated carbonyl compound (3), as shown in Eq. 2.1.[1]

H $\xrightarrow{SeO_2}$ OH and/or O   (2.1)

1 2 3

The mechanism of the reaction has been studied extensively.[1] According to Sharpless[2,3] it involves an initial ene addition of selenium dioxide to the less hindered side of the olefin, followed by a [2,3]-sigmatropic shift of the resulting allylic seleninic acid (**4**), finally leading to an unstable selenium(II) ester (**5**), which readily cleaves to form the allylic alcohol (**2**), as shown in Eq. 2.2. The corresponding carbonyl compound (**3**) results from further oxidation by selenium dioxide. Complementary to this *concerted* mechanism is a pathway involving ionic intermediates, which is favored in solvents such as *t*-butyl alcohol.[3,4] A dissociation–recombination pathway with radical intermediates was also proposed, which is favored with endocyclic olefins.[3]

H O Se O — ene → O Se−OH — [2,3]-shift → OSeOH

1 4 5

Hydrolysis

O — SeO_2 ← OH (2.2)

3 2

The remarkable regioselectivity of this oxygenation has been utilized in many natural product syntheses. Thus, trisubstituted olefins of the general type **6** (Eq. 2.3) are selectively oxygenated at the methyl group *trans* to the alkyl group R, producing the *trans*-allylic alcohol **7** or the aldehyde **8**.[6d]

R, Me, Me — SeO_2 → R, Me, CH_2OH and/or R, Me, CHO (2.3)

6 7 8

Usually the reaction gives a mixture of **7** and **8**, which is treated with sodium borohydride or lithium aluminum hydride if the alcohol (**7**) is desired, or with manganese dioxide if the desired product is the aldehyde (**8**). A typical example is the conversion of β-santalene (**9**) to E-β-santalol (**10**), shown in Eq. 2.4.[5]

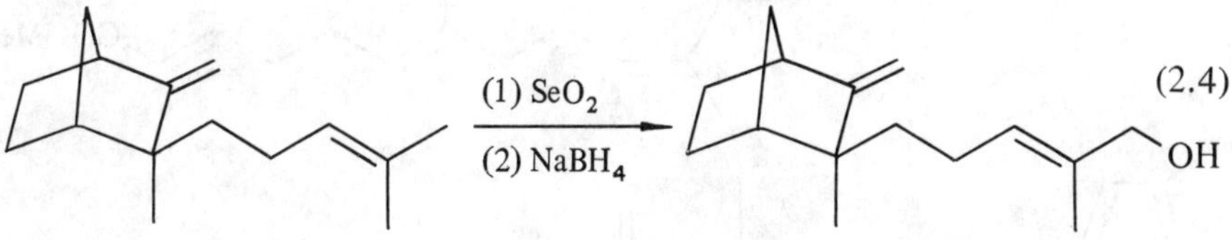

(1) SeO_2
(2) $NaBH_4$

OH

9 **10** (2.4)

This type of operation was crucial in converting intermediate **11** to the chemotactic hormone (−)-sirenin (**12**, Eq. 2.5).[6]

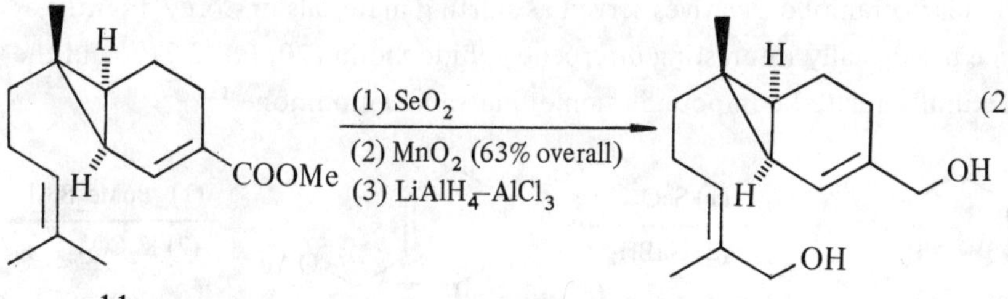

H H COOMe

(1) SeO_2
(2) MnO_2 (63% overall)
(3) $LiAlH_4$–$AlCl_3$

H H OH OH

11 **12** (2.5)

The use of geraniol derivatives (**13**) in this reaction gives access to useful synthetic intermediates such as **14** and **15** (Eq. 2.6).

OR X OR

13

14: X = CHO
15: X = CH_2OH (2.6)
16: X = CH_2Br

Aldehyde **14**, for example (R = acetyl[7] or mesitoyl[8]), was used in the total synthesis of the nor-sesquiterpene gyrinidal (**17**)[7,8] and in a synthesis of (±)-C_{17}-cecropia juvenile hormone (**18**) via bromide **16** (R = THP).[9] The same bromide (**16**, R = THP), prepared from alcohol **15** (R = THP) in a different way, was used in the total synthesis of pleraplysillin-2 (**19**), a sponge metabolite.[10]

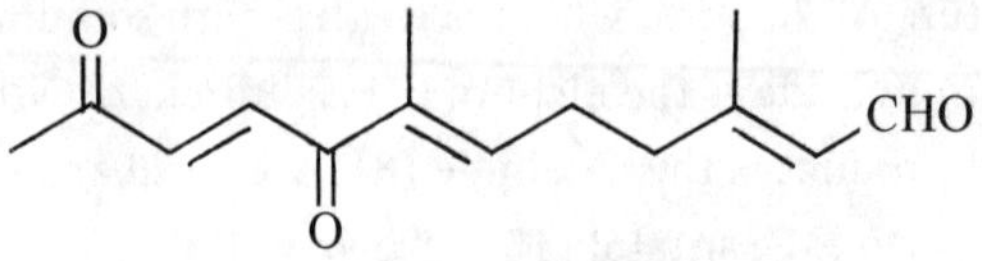

CHO

17

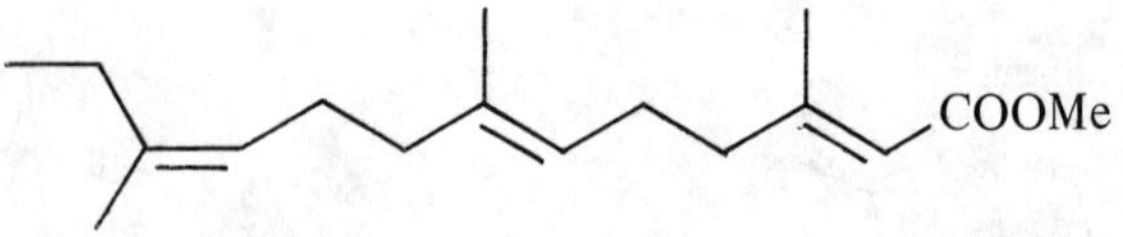

COOMe

18

O

O

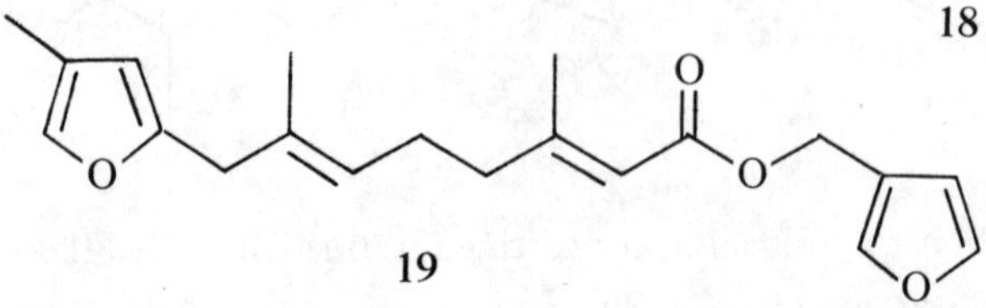

19

(2.6)

Similar geraniol derivatives served as starting materials in Corey's synthesis of the biologically interesting diterpene aphidocholin (20, Eq. 2.7)[11] and the structurally related compounds stemodine and stemodinone.[12]

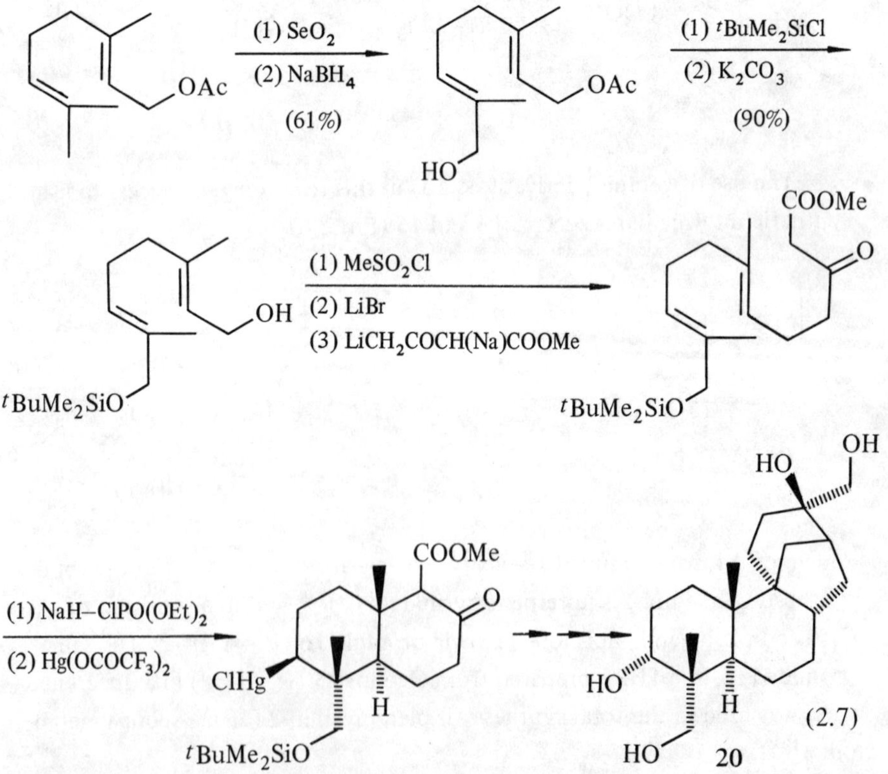

OAc (1) SeO_2 (2) $NaBH_4$ (61%) OAc HO (1) $^tBuMe_2SiCl$ (2) K_2CO_3 (90%)

OH (1) $MeSO_2Cl$ (2) LiBr (3) $LiCH_2COCH(Na)COOMe$ $^tBuMe_2SiO$ COOMe O $^tBuMe_2SiO$

(1) NaH–$ClPO(OEt)_2$ (2) $Hg(OCOCF_3)_2$ COOMe O ClHg H $^tBuMe_2SiO$ HO OH HO H HO 20

(2.7)

Oxidation of citronellal ethylene ketal (**21**) with selenium dioxide gave aldehyde **22**, which on treatment with acid, cyclized to irridodial (**23**), a precursor of iso-irridomyrmecin (**24**, Eq. 2.8).[13]

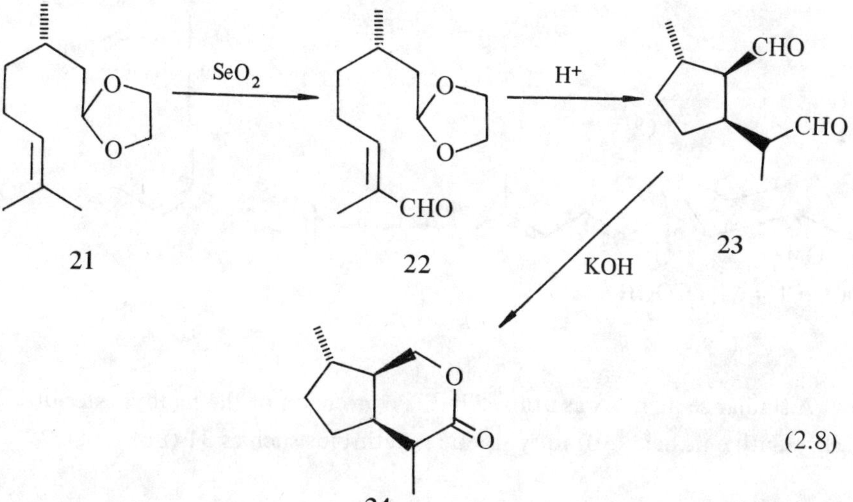

SeO_2

H^+

CHO

CHO

CHO

21

22

23

KOH

O

O

(2.8)

24

In an elegant synthesis of α-sinensal (**27**), myrcene (**25**) was oxidized with selenium dioxide to alcohol **26**, which was transformed to **27** by transetherification with the appropriate vinyl ether, followed by two consecutive Cope rearrangements (Eq. 2.9).[14]

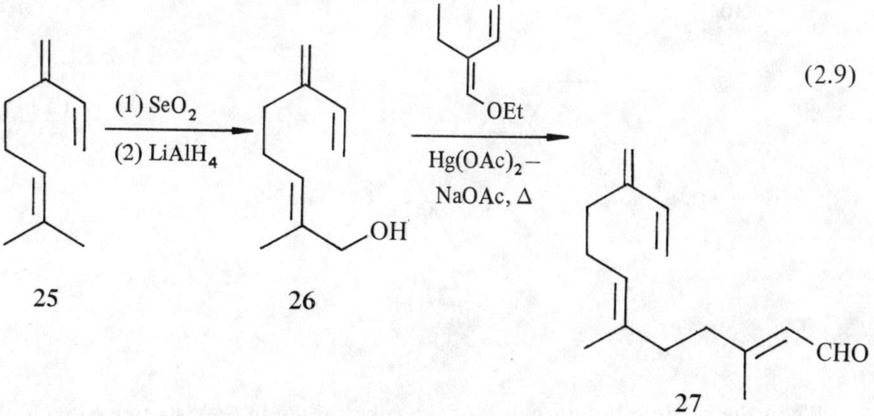

(2.9)

(1) SeO_2

(2) $LiAlH_4$

OEt

$Hg(OAc)_2$–

NaOAc, Δ

OH

25

26

CHO

27

Allylic oxidation with selenium dioxide was also used to prepare **28**, which served as an intermediate in the synthesis of the antibiotic fumagillin (**29**, Eq. 2.10).[15]

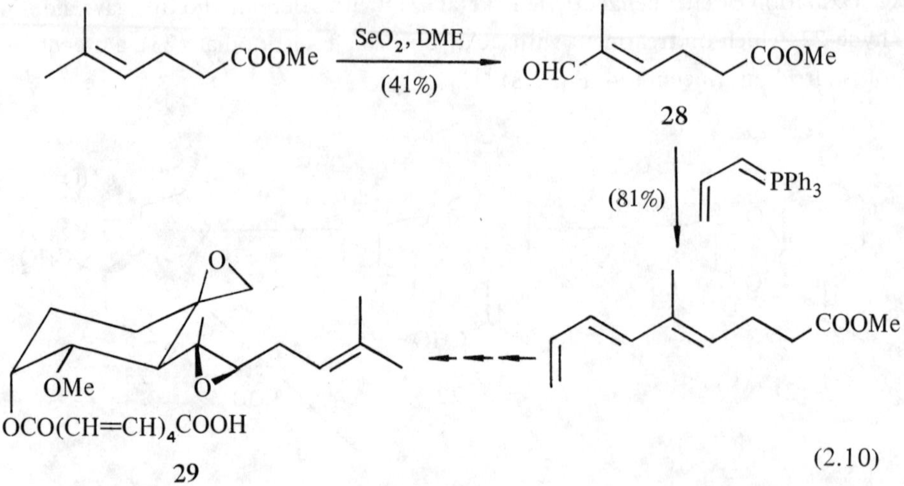

(2.10)

A similar sequence was utilized in the conversion of the methyl ester of chrysanthemic acid (**30**) to synthetic pyrethroids such as **31** (Eq. 2.11).[16]

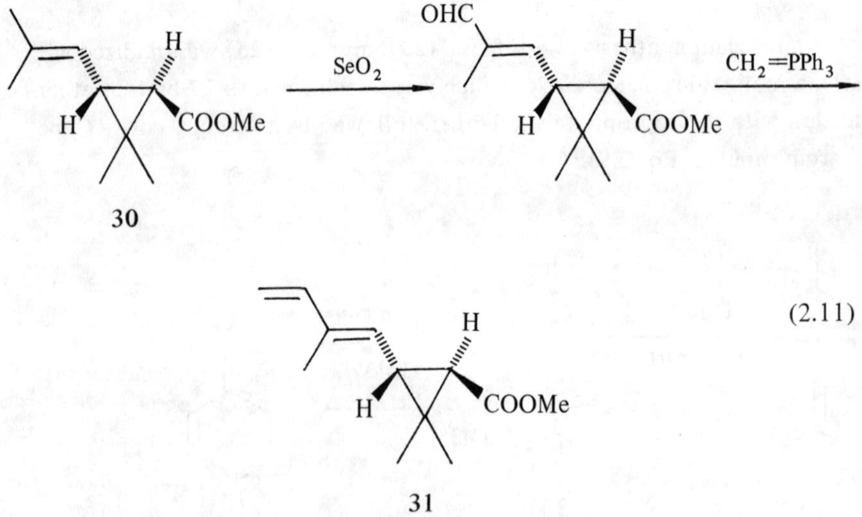

(2.11)

When the olefinic functionality of the type shown in **6** exists twice in a molecule, the reaction can occur in both places, as exemplified in the oxidation of diene **32** to form the all-*trans* diol (**33**), which was elaborated further to squalene (**34**) by a three-step sequence (Eq. 2.12).[17]

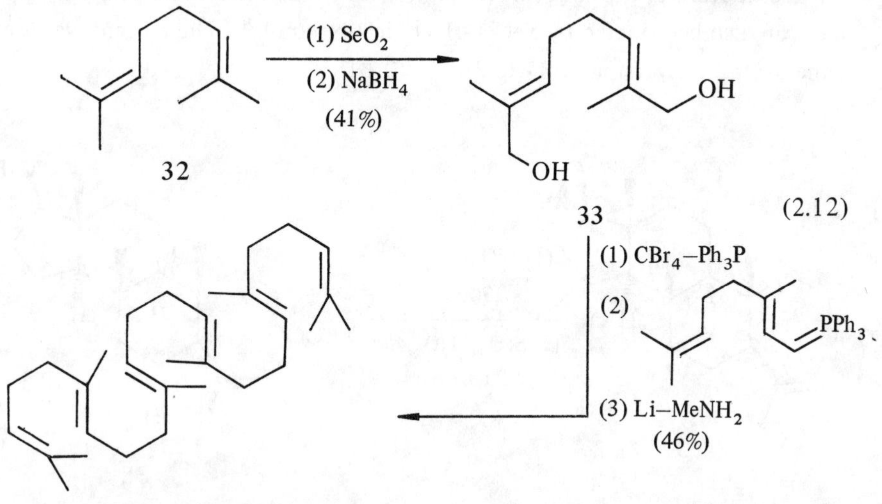

(1) SeO_2
(2) $NaBH_4$
(41%)

32 33

(2.12)

(1) CBr_4–Ph_3P
(2) PPh$_3$
(3) Li–$MeNH_2$
(46%)

34

Further oxidation of the α,β-unsaturated aldehydes (**8**) according to Corey's procedure[18] allows the preparation of the corresponding carboxylic acid methyl esters. For example, lanosterol acetate (**35**) gives, after hydrolysis, ganoderic acid Z (**36**, Eq. 2.13).[19]

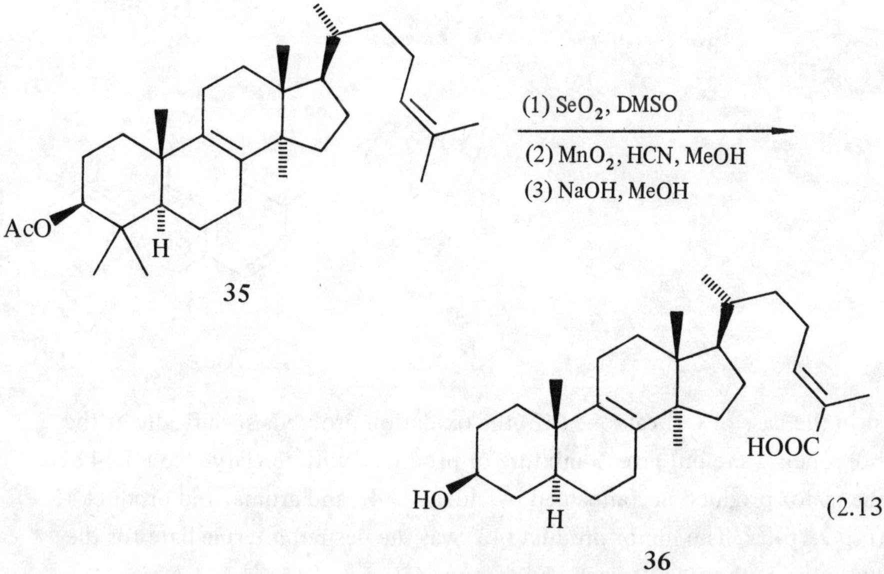

AcO H

35

(1) SeO_2, DMSO
(2) MnO_2, HCN, MeOH
(3) NaOH, MeOH

HOOC
HO H

(2.13)

36

The oxidation of α,β-unsaturated aldehydes to α,β-unsaturated carboxylic acids can also be effected by SeO_2–H_2O_2 in *t*-butanol.[20] The reagent was used in the overall conversion of **37** to **38** (Eq. 2.14).[21]

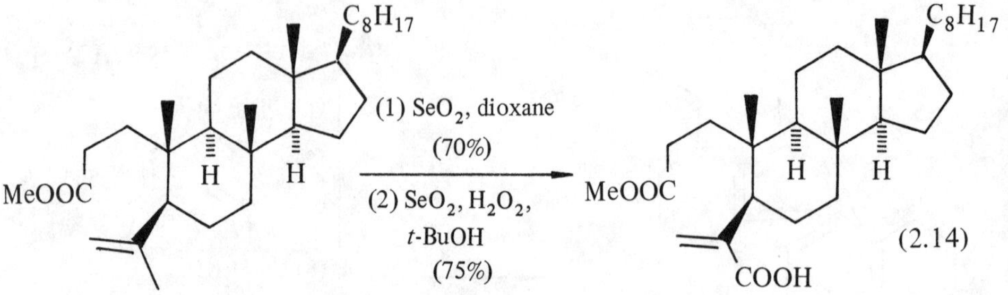

C_8H_{17} MeOOC H H (1) SeO_2, dioxane (70%) (2) SeO_2, H_2O_2, *t*-BuOH (75%) C_8H_{17} MeOOC H H COOH (2.14)

37 **38**

Another example of regioselectivity in the reaction of selenium dioxide with olefins is the oxidation of β-damascenone (**39**) to give **40** as the major product, together with the rearranged aromatic product (**41**, Eq. 2.15).[22]

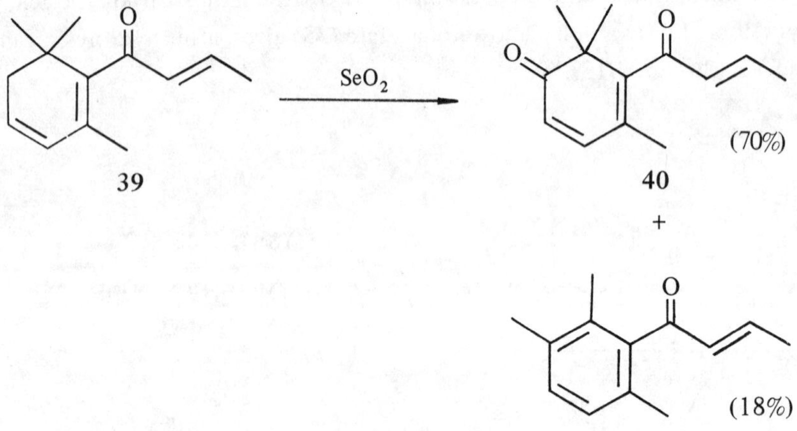

O SeO_2 O O (70%)

39 **40**

\+

O (18%)

41

(2.15)

In the case of (+)-carvone (**42**) the oxidation proceeds specifically at the isopropenyl site but gives a mixture of products, with tertiary alcohol **44** as the major product, accompanied by aldehyde **43** and aromatized product **45** (Eq. 2.16).[22] The minor product (**43**) was the desired intermediate for the synthesis of the sesquiterpene bilobanone (**46**, Eq. 2.16).[23]

SeO_2

OHC

(10%) (70%)

OH

42 43 44

OH

43

(51%) *i*-PrC≡CMgBr

(8%)

45

O O

OH

(1) H_2SO_4

(2) $HgSO_4$

O

46 (2.16)

Oxidation of (+)-*trans*-verbenyl acetate (**47**) with selenium dioxide in dioxane gave aldehyde **48**, thereby opening the way for a chain extension, as shown in Eq. 2.17.[24] Acetates **47** and **49** were found to be mimics of the sex pheromone of the American cockroach.[24]

CHO

(1) $NaBH_4$

SeO_2, dioxane (2) CCl_4, PPh_3

(53%) (3) KOH

OAc OAc OAc

(4) Me_2CuLi

47 48 (5) Ac_2O, pyr 49

(75% overall) (2.17)

Also, Kishi[25] has used a selenium dioxide oxidation of this type to convert **50** to alcohol **51**, which is an intermediate in his synthesis of tetrodotoxin (**52**, Eq. 2.18).[25]

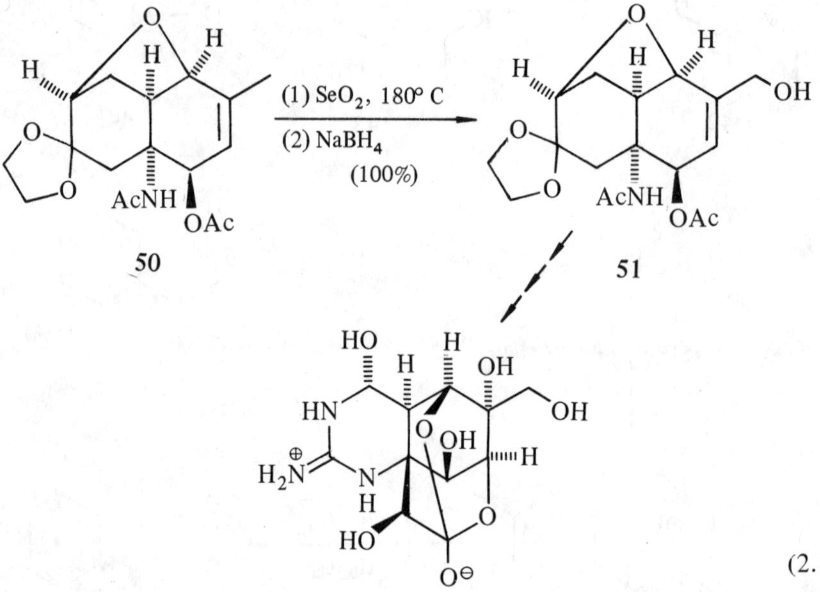

(1) SeO_2, 180° C
(2) $NaBH_4$
(100%)

50 **51**

52

(2.18)

If the substrate contains a carboxylic acid or ester group in the neighborhood of the allylic oxidation, then the product is a lactone (**53**, Eq. 2.19).

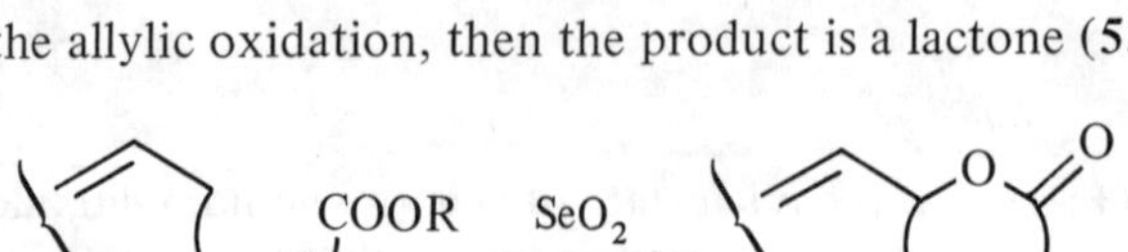

COOR SeO_2 (2.19)

R = H, Me or Et **53**

This type of reaction was used in the synthesis of (±)-santonin (**54**, Eq. 2.20)[26] and (±)-clovene (**55**, Eq. 2.21).[27]

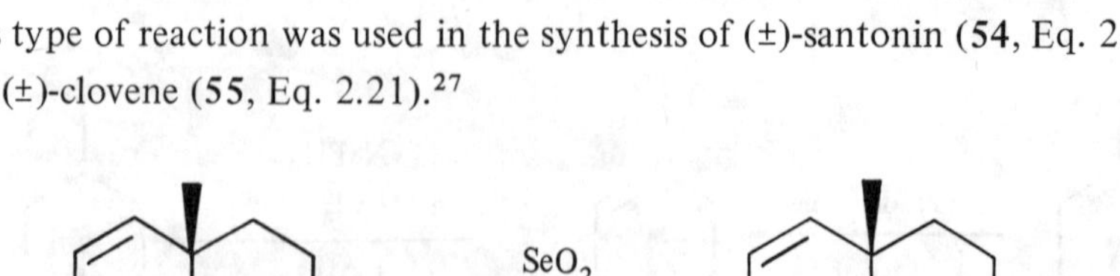

SeO_2

COOH (2.20)

54

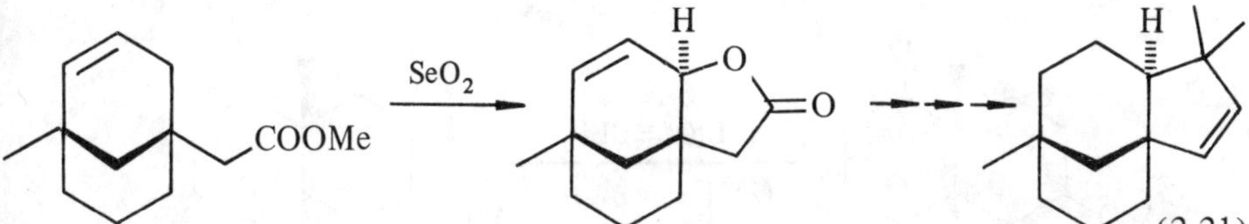

SeO_2

55

(2.21)

Similar methodology is applicable for the synthesis of butenolides (**56**, Eq. 2.22).

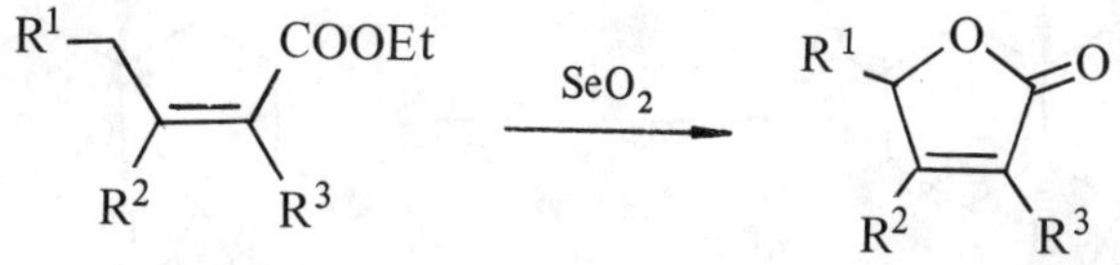

SeO_2

56

(2.22)

Sondheimer has used this chemistry in the synthesis of (+)-digitoxigenin (**57**, Eq. 2.23).[28]

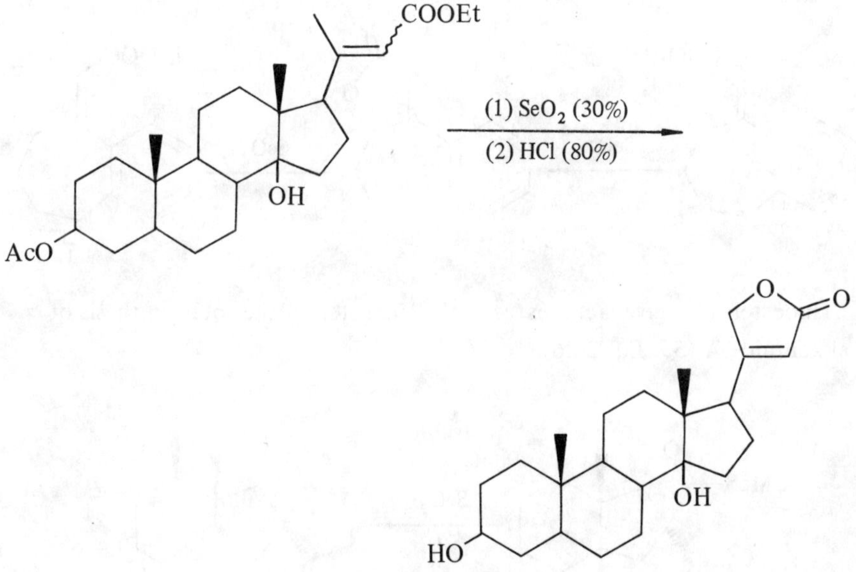

(1) SeO_2 (30%)
(2) HCl (80%)

57

(2.23)

Further elaboration of the butenolides synthesized by this method has led to other useful intermediates, such as **58** (Eq. 2.24).[29]

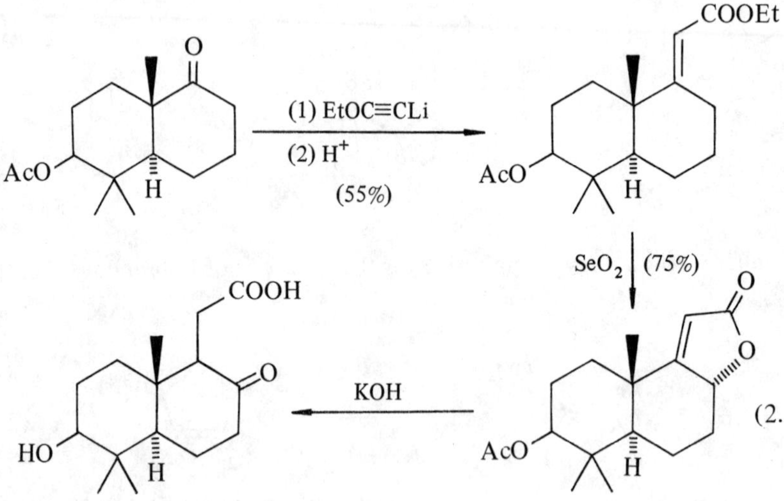

(1) EtOC≡CLi
(2) H^+
(55%)

SeO_2 (75%)

KOH

58

(2.24)

Interestingly, it was found recently that both the *cis* and the *trans* α,β-unsaturated carboxylic esters give the same butenolide with equal ease (Eq. 2.25).[30]

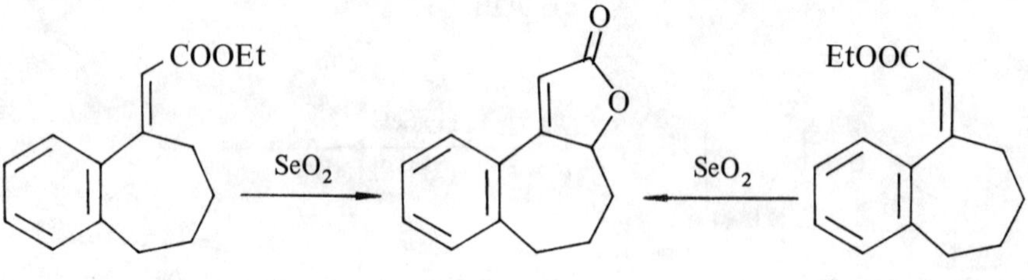

SeO_2 SeO_2

(2.25)

Thioesters also give lactones, as in the final step of the total synthesis of bakkenolide A (**59**, Eq. 2.26).[31]

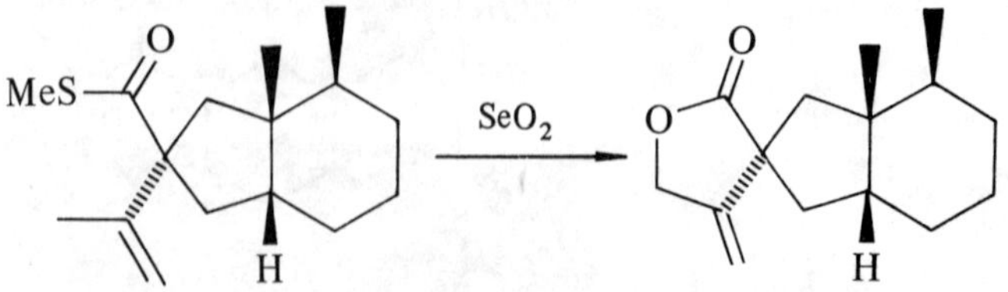

SeO_2

59

(2.26)

Oxidation of disubstituted olefins with selenium dioxide can be accompanied by allylic rearrangement (Eq. 2.27).

SeO_2

OH

(2.27)

This type of rearrangement was desired in the overall regiospecific introduction of a carbonyl group in one of the two olefinic carbons of **60** to give **61**, a useful intermediate for alkaloid synthesis (Eq. 2.28).[32]

(1) S O_2 CH_2COCl

(2) Δ

N H H O

60

(26%) (1) SeO_2, AcOH (2) KOH (3) $H_2CrO_4 \cdot pyr$

H_2/cat

61

(2.28)

The rearranged product was also obtained during the oxidation of **62** to give **63**, which was used in a total synthesis of the antitumor alkaloid camptothecin (**64**, Eq. 2.29).[33]

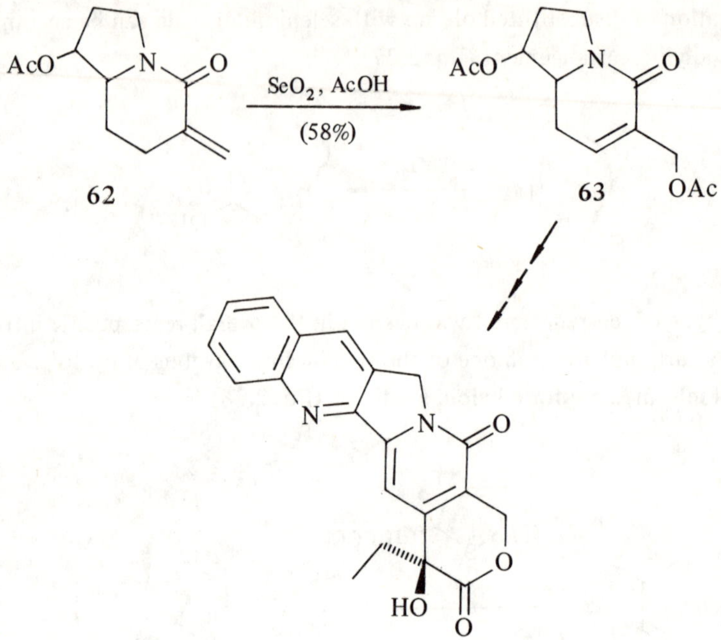

AcO 62 $\xrightarrow[(58\%)]{SeO_2,\ AcOH}$ AcO 63 OAc

64

(2.29)

A serious drawback of the selenium dioxide oxidation of olefins is the formation of selenium by-products that are hard to remove from the reaction mixture. In order to avoid this problem, Sharpless has recently introduced a modification in which selenium dioxide is used catalytically or stoichiometrically in the presence of *t*-butyl hydroperoxide (*t*-BuOOH), and in some cases, with acid catalysis from salicylic acid.[34] These conditions are milder and are usually more effective. For example, β-pinene (65) gives *trans*-pinocarveol (**66**) in 86% yield (Eq. 2.30).[34] Using hydrogen peroxide in *t*-butanol instead of *t*-butyl hydroperoxide, although successful with less reactive olefins,[34] gives the same product (**61**) with β-pinene, but in lower yield.[35]

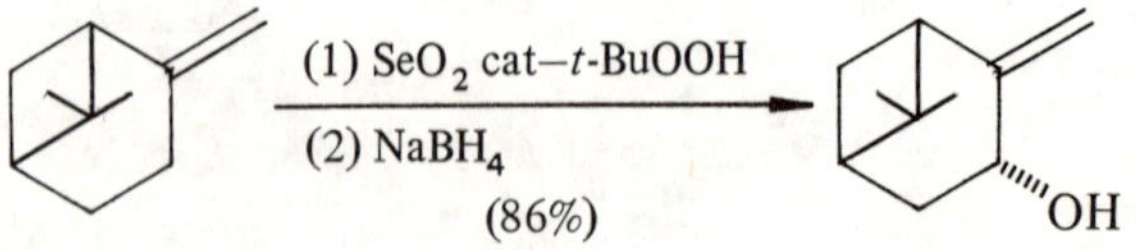

65 $\xrightarrow[(2)\ NaBH_4\ (86\%)]{(1)\ SeO_2\ cat\text{–}t\text{-}BuOOH}$ 66 (OH) (2.30)

The Sharpless procedure for allylic hydroxylation,[34] besides solving the problem of malodorous and poisonous selenium by-products, avoids several side reactions of selenium dioxide such as rearrangements (Eq. 2.27), dehydrations and dehydrogenations (Chapter 3). Furthermore, being a milder reaction, it has enhanced regioselectivity and stereoselectivity. All of these advantages, combined with its high efficiency, make it the method of choice for transformations of the general type shown in Eq. 2.1.

The high regioselectivity of the procedure is indicated in the conversion of **67** to **68** (Eq. 2.31)[36] and in the oxidation of epitulipinolide (**69**, Eq. 2.32)[37] and humulene-9,10-epoxide (**70**, Eq. 2.33).[38] Another example is found in the synthesis of the hydroxy derivative of vitamin D (**71**, Eq. 2.34).[39] Some mechanistic explanations for the observed selectivity were also offered.[37, 39] A similar reaction was recently employed by Trost during studies on the stereospecific introduction of steroid side chains (Eq. 2.34a).[84] The same transformation with only selenium dioxide,[84b] was less efficient.

67 →[(1) SeO_2 cat–*t*-BuOOH; (2) PCC; (50%)] **68** (2.31)

69 →[SeO_2 cat–*t*-BuOOH; (90%)] (2.32)

70 →[SeO_2–*t*-BuOOH; (45%)] (2.33)

(1) TsCl, pyr

(2) MeOH, NaOAc

SeO_2–t-BuOOH (50%)

(1) AcOH

(2) NaOH

(2.34)

71

SeO_2–t-BuOOH (80%)

(2.34a)

Further applications of the Sharpless method are found in the synthesis of the marine natural product mokupalide (**72**, Eq. 2.35),[40] of the diterpene trihydroxy-decipiadiene (**73**, Eq. 2.36)[41] and the sesquiterpene coriamyrtin (**74**, Eq. 2.37).[42]

SeO_2–t-BuOOH (38%)

COOMe

HO COOMe

72

(2.35)

SeO_2 cat–t-BuOOH (50%)

H H H H OMe

OH

(1) HCl
(2) $Ph_3P{=}CMeCOOEt$
(3) DIBAL

73

(2.36)

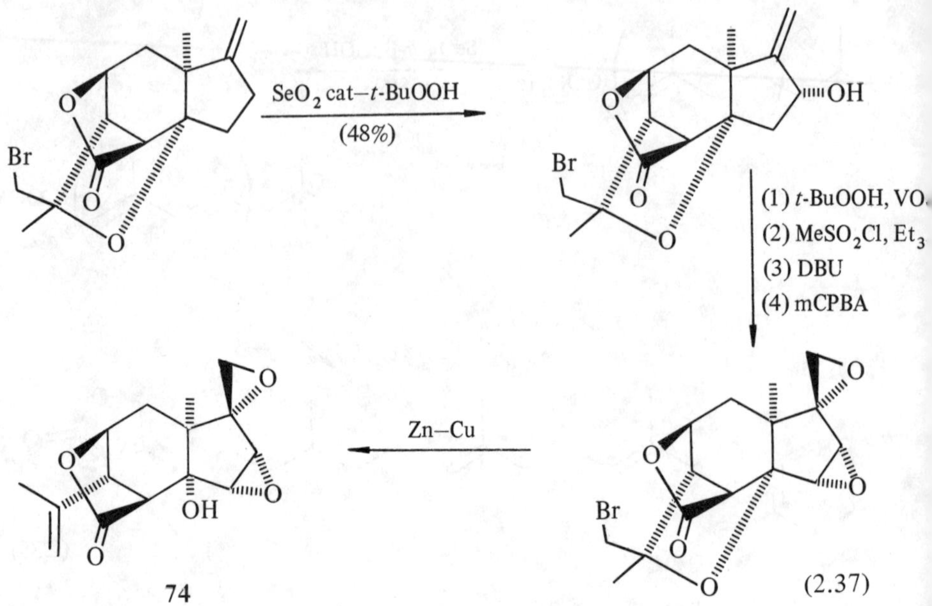

SeO2 cat–t-BuOOH
(48%)
(1) t-BuOOH, VO
(2) MeSO2Cl, Et3
(3) DBU
(4) mCPBA
Zn–Cu
Br
OH

74

(2.37)

In some cases a mixture of isomeric olefins is subjected to Sharpless oxidation and the desired allylic alcohol is separated from the oxidation mixture. Two examples are shown in Eq. 2.38[43] and Eq. 2.39.[44]

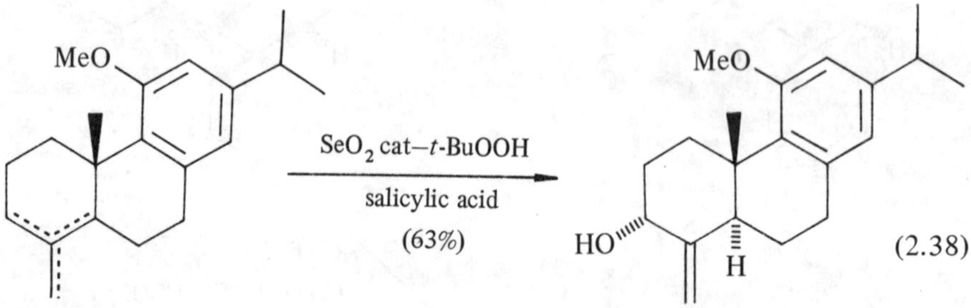

MeO
SeO2 cat–t-BuOOH
salicylic acid
(63%)
HO
H

(2.38)

(2.39)

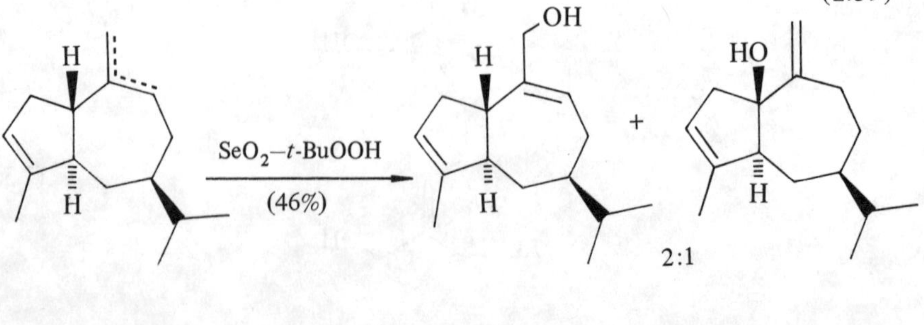

H
SeO2–t-BuOOH
(46%)
OH
HO
+

2:1

A modification[45] of the Sharpless procedure, involving the use of selenium dioxide supported on silica gel, allows the selective oxidation of allylic methyl groups attached to medium size ring compounds without any side reactions, such as epoxidation of olefinic bonds or endocyclic oxidation. Thus, humulene (**75**) gives a mixture of the mono- and dioxygenated products (Eq. 2.40),[45] while caryophylene (**76**) forms the corresponding allylic alcohol **77** in which the internal olefinic bond is isomerized to the more stable *cis* form (Eq. 2.41).[45]

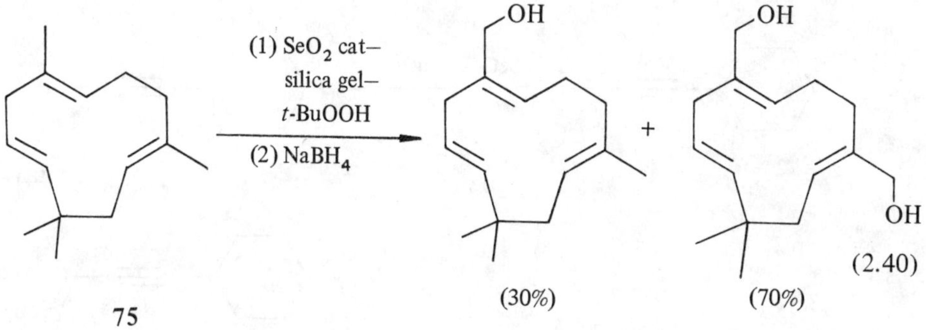

(1) SeO_2 cat–silica gel–*t*-BuOOH
(2) $NaBH_4$

75 → (30%) + (70%) (2.40)

(1) SeO_2 cat–silica gel–*t*-BuOOH
(2) $NaBH_4$
(90%)

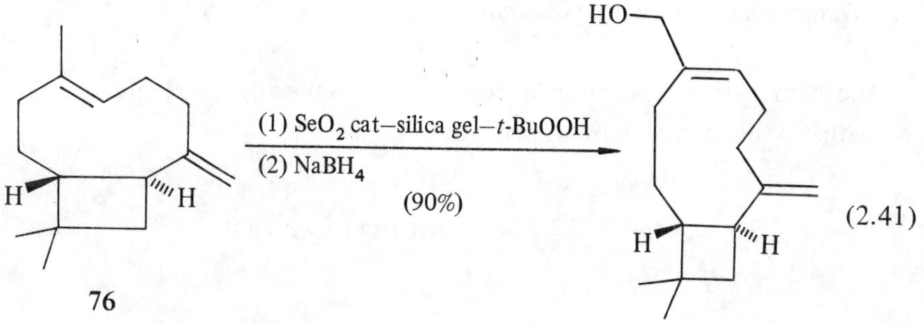

76 → **77** (2.41)

2.2 α-Oxygenation of alkynes

Examples of α-oxygenation of alkynes similar to the allylic oxidation of olefins with selenium dioxide, are relatively rare. However, Sharpless has shown that his SeO_2–*t*-BuOOH procedure can also be applied to acetylenic compounds.[46] The only difference is that alkynes tend to undergo oxygenation on both α-sides of the acetylenic bond (Eq. 2.42). Furthermore, on suitable substrates, dehydration of the resulting acetylenic alcohol can lead to enynones (Eq. 2.44).

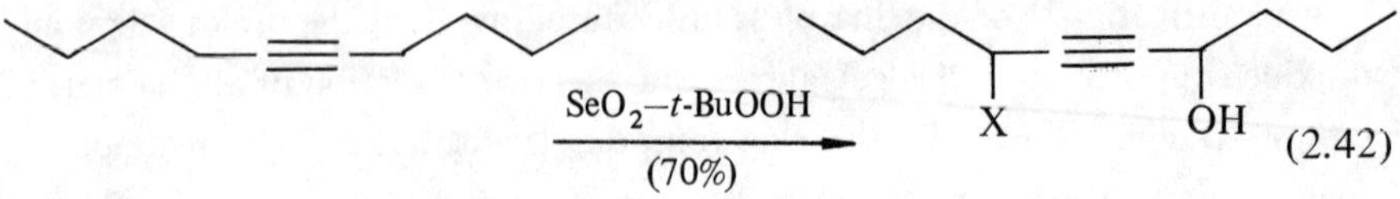

$$\xrightarrow[\text{(70\%)}]{SeO_2\text{–}t\text{-BuOOH}} \quad \text{X, OH} \qquad (2.42)$$

X = H or OH (1 : 2)

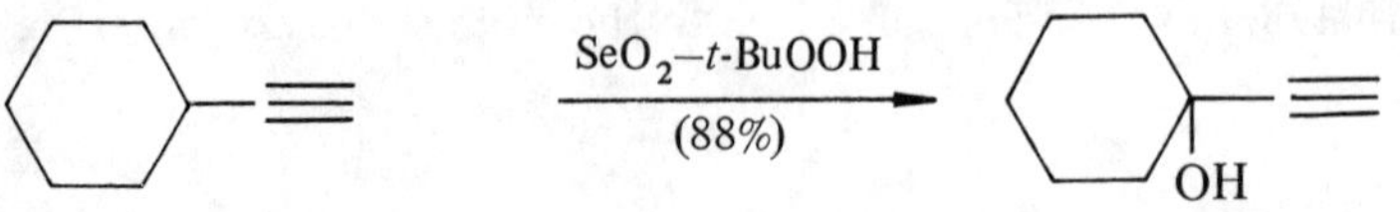

$$\xrightarrow[\text{(88\%)}]{SeO_2\text{–}t\text{-BuOOH}} \quad \text{OH} \qquad (2.43)$$

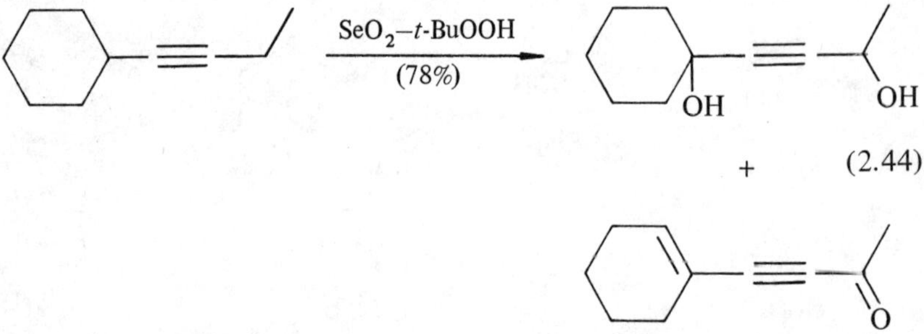

$$\xrightarrow[\text{(78\%)}]{SeO_2\text{–}t\text{-BuOOH}} \quad \text{OH} \quad \text{OH} \quad + \qquad (2.44)$$

O

2.3 *Oxidation of aromatic side chains*

Another reaction of selenium dioxide related to the allylic oxidation of olefins is the oxidation of aromatic side chains (Eq. 2.45).

$$\text{ArMe} \xrightarrow{SeO_2} \text{ArCHO or ArCOOH} \qquad (2.45)$$

Unfortunately, the reaction is not synthetically useful for carbocyclic aromatic systems, due to the severe conditions required. An example is shown in Eq. 2.46.[47]

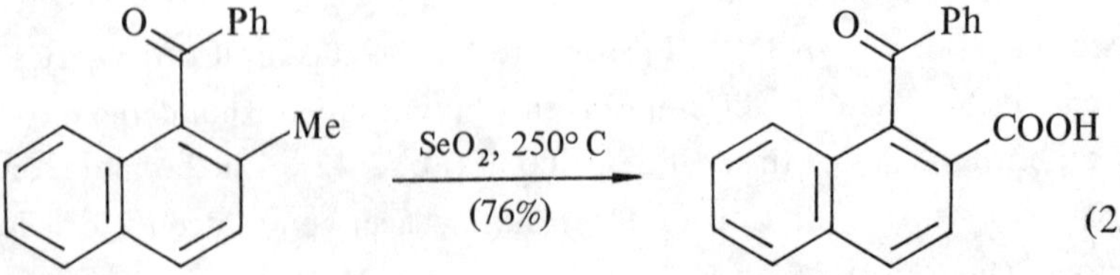

O, Ph, Me

$$\xrightarrow[\text{(76\%)}]{SeO_2,\ 250°\ C}$$

O, Ph, COOH (2.46)

With heterocyclic aromatics, this oxidation can be performed under relatively mild conditions and with remarkable regioselectivity (due to special position activation) as shown in the case of trimethylquinoline (**78**, Eq. 2.47).[48] This type of reaction was applied to the synthesis of the antitumor antibiotic streptonigrin.[85]

SeO_2, EtOH, Δ (82%)

78

(2.47)

Under different conditions, the product can be the carboxylic acid, as shown in Eq. 2.48.

SeO_2, pyr, 115° C (79%)

(2.48)

Selective oxidation of the methyl group in the pyridine ring of **79** with selenium dioxide was a key step in the synthesis of nybomycin (**80**, Eq. 2.49).[50] The methyl group in the pyridone ring required much higher temperatures to be oxidized.

SeO_2, dioxane, Δ (53%)

79

(1) $NaBH_4$
(2) AcOH
(3) Me_2SO_4
(4) $K_3Fe(CN)_6$, KOH
(5) KOH excess

80

(2.49)

By using Sharpless' conditions, Cook was able to convert **81** to **82** (Eq. 2.50).

SeO_2–*t*-BuOOH
(50%)

81 82 (2.50)

Oxygen heterocycles behave similarly. For example, methylcoumarin derivative **83** was oxidized by Büchi to the corresponding aldehyde (**84**), which was elaborated to aflatoxin B_1 (**85**, Eq. 2.52).[52]

SeO_2, xylene, Δ
(93%)

83 84

(80%) Zn, AcOH, Δ

85 (2.51)

Methyl derivatives of 2-pyrones can also be oxidized in the same fashion. Thus, **86** was oxidized to aldehyde **87** which is a useful intermediate for the synthesis of naturally occurring 2-pyrones, such as yangonin (**88**, Eq. 2.52).[53]

86 —(SeO_2, dioxane, Δ; (75%))→ 87 —(p-$MeOC_6H_4CH{=}PPh_3$; (93%))→ 88 (2.52)

A milder method for oxidations of the type shown in Eq. 2.45 involves the use of benzeneseleninic anhydride, PhSe(O)OSe(O)Ph, a reagent recently introduced by Barton.[54] The method works with a variety of carbocyclic and heterocyclic aromatic compounds, and with certain activated rings it proceeds with competing incorporation of PhSe groups on the aromatic ring. Two typical examples are shown in Eq. 2.53[54] and Eq. 2.54.[54]

p-xylene —(PhSe(O)OSe(O)Ph, 120° C; (62%))→ p-tolualdehyde (2.53)

2,6-dimethoxytoluene —(PhSe(O)OSe(O)Ph, 110° C; (85%))→ 2,6-dimethoxy-3-(phenylseleno)toluene (2.54)

2.4 α-Oxygenation of carbonyl compounds

The one-step oxidation of ketones (**89**) to α-dicarbonyl compounds (**90**) with selenium dioxide is a very useful synthetic transformation. The unique ability of selenium dioxide to effect this transformation has been known for a long time and has been used on numerous occasions.[1]

SeO_2

89 **90**

(2.55)

The mechanism of the reaction has created some controversy. Corey proposed the mechanism shown in Eq. 2.56,[55] while more recently, Sharpless provided evidence for an alternative mechanism, via organoselenium intermediates, indicated in Eq. 2.57.[56]

OH SeO_2 O Se OH O O OSeOH

Base (2.56)

SeO_2

OH Se O Pummerer-type rearrangement O Se O $-Se^0$ O O

(2.57)

A typical example is the oxidation of camphor (**91**) to camphorquinone (**92**) in high yield.[57]

SeO_2, Ac_2O (96%)

91 **92**

(2.58)

This type of oxidation was recently used in a synthesis of phytohormone analogs, outlined in Eq. 2.59.[58]

SeO_2

(1) Zn–AcOH
(2) $NaBH_4$
(3) H^+, acetone

(1) Na_2CO_3
(2) $MeSO_2Cl$–pyr, Δ

HIO_4

H^+

(2.59)

The formation of a 1,2-dione system (**90**) allows the insertion of an oxygen atom between the two carbonyls with hydrogen peroxide.[57] For example, **93** was converted to **94** (Eq. 2.60).[59]

(1) SeO_2, AcOH (83%)
(2) H_2O_2 (86%)

93 **94**

(2.60)

Appropriately substituted diones may favor the enolization of one of the carbonyls, permitting various selective transformations. This was the case with **96**, the selenium dioxide oxidation product of **95**, which upon treatment with hydrazine hydrate at 155° C produced the alkaloid lycopodine (**97**), together with **98**, another naturally occurring alkaloid (Eq. 2.61).[60]

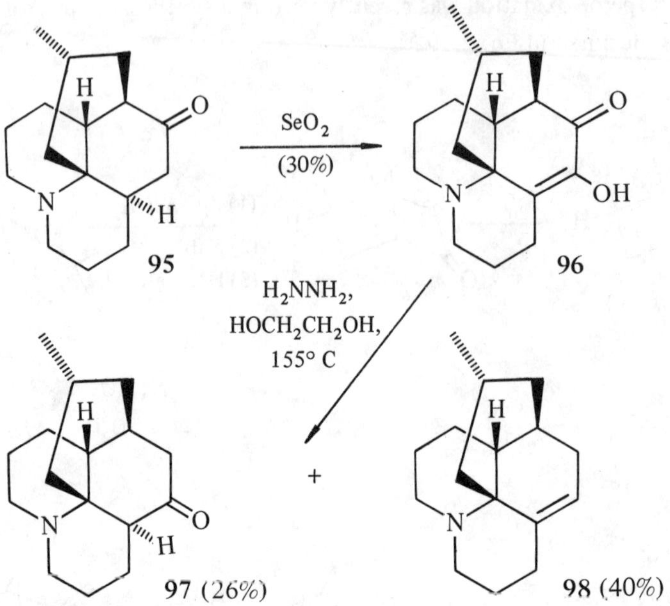

SeO$_2$ (30%)

95 → 96

H_2NNH_2, $HOCH_2CH_2OH$, 155° C

97 (26%) + 98 (40%)

(2.61)

In Woodward's synthesis of strychnine (**101**), one of the earliest masterworks of natural products synthesis, selenium dioxide oxidation of methyl ketone **99** led selectively to the corresponding glyoxal, which upon epimerization to the *cis* isomer, cyclization and further oxidation gave in one step α-lactam **100**, which was further elaborated to strychnine (**101**, Eq. 2.62).[61]

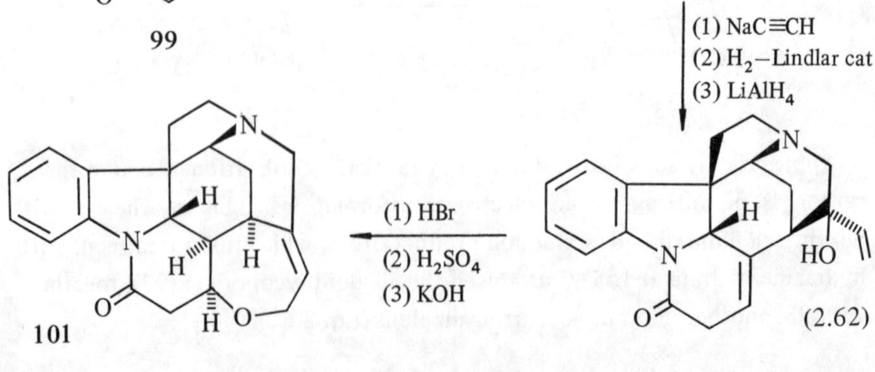

SeO$_2$ (12%)

99 → 100

(1) NaC≡CH
(2) H_2–Lindlar cat
(3) $LiAlH_4$

(1) HBr
(2) H_2SO_4
(3) KOH

101

(2.62)

Oxidation of methyl ketones in pyridine at 100° C gives the corresponding ketocarboxylic acids. The reaction was used in the conversion of **102** to **104** via **103** in the synthesis of the oxygen analog of serotonin, oxaserotonin (**105**, Eq. 2.63).[62]

PhCH$_2$O … COOEt **102** —[SeO_2, pyr, 100° C (82%)]→ PhCH$_2$O … COOH, COOEt **103** —[(59%) K_2CO_3, Δ]→ PhCH$_2$O … COOH, COOH **104** → → → HO … NH$_2$ **105**

(2.63)

With suitable dicarbonyl substrates, treatment with selenium dioxide can lead to oxidative cleavage, as in the conversion of **106** to **107** (Eq. 2.64).[63]

106 (OH, O, NH, O) —[SeO_2, AcOH (89%)]→ **107** (O, O, NH, O) (2.64)

N-Alkyl derivatives of **107**, such as **109**, have also been prepared by selenium dioxide oxidation of the appropriate *N*-alkyl homophthalimide (**108**, Eq. 2.65).[64] Condensation of **109** with 5,6-diamino-1,3-dimethyluracil (**110**) gave the novel heterocyclic system **111** (Eq. 2.65).[64]

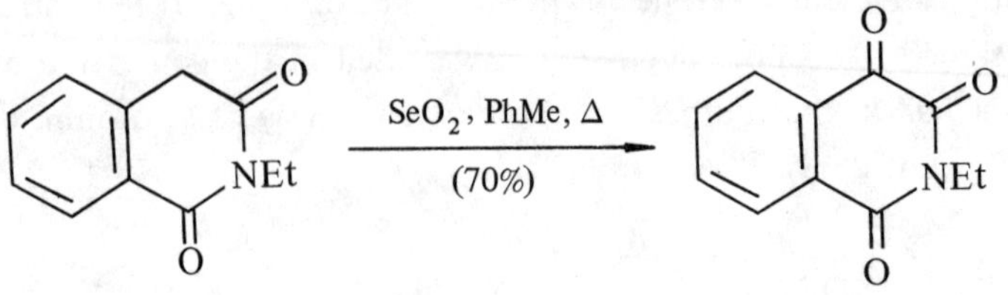

SeO_2, PhMe, Δ (70%)

108 → 109

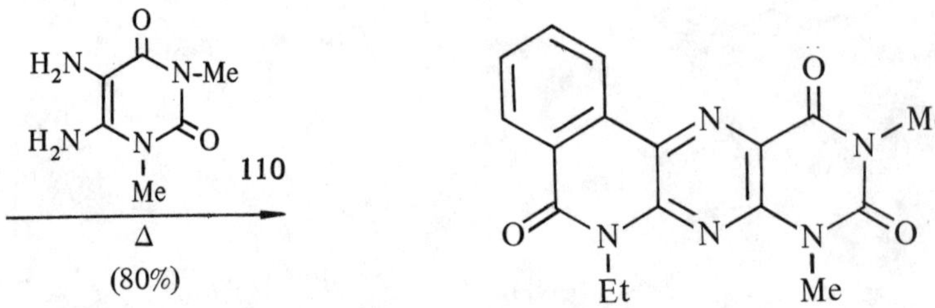

110, Δ (80%)

111

(2.65)

Introduction of a tertiary hydroxyl group at a position α- to a ketone cannot be readily achieved with selenium dioxide. However, α-hydroxy ketones can be prepared by reacting phenols or sodium phenoxides with benzeneseleninic anhydride, PhSe(O)OSe(O)Ph, a reagent recently studied by Barton and co-workers.[65] Thus, upon treatment with this reagent, **112** gave **113**, which has a substitution pattern similar to that of the *A* ring of tetracyclin (**114**, Eq. 2.66).[65]

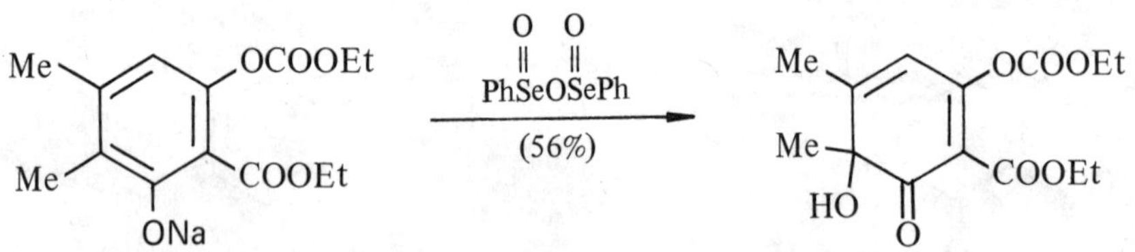

PhSeOSePh (56%)

112 → 113

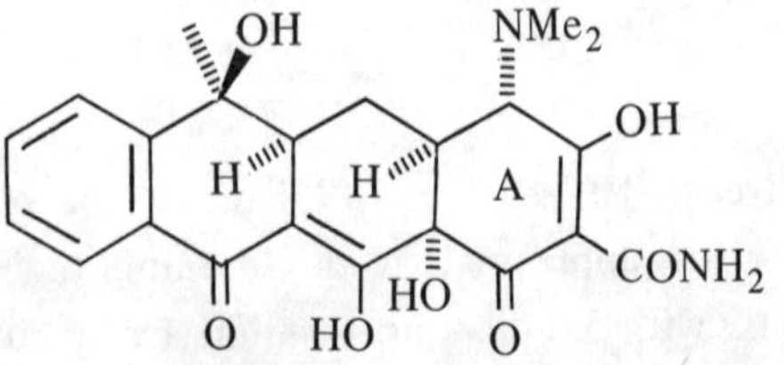

114

(2.66)

Using this method, Jeffs converted naphthol **115** to naphthalenone **116**, a compound isolated from cotton and showing chemotactic activity (Eq. 2.67).[66]

PhSeOSePh (O O), (85%)

115 → **116** (2.67)

The reaction is presumed by Barton to proceed via a [2,3]-sigmatropic rearrangement of an intermediate seleninyl ester (**118**, Eq. 2.68).[65] In accordance with this mechanism, *a*-unsubstituted phenols (**117a**, R^2 = H) produce the corresponding *o*-quinones (**121**), which arise from **119a** (R^2 = H) by further oxidation, instead of the *a*-hydroxy ketones (**120**) to be expected from *a*-substituted phenols (**117b**, $R^2 \neq$ H Eq. 2.68).[65,67] Possible involvement of a pyrocatechol intermediate (**122**) in the transformation of **119a** (R^2 = H) to **121** has not been confirmed.[67]

117 → (PhSeOSePh) → **118** → **119**; **119** → ($R^2 \neq$ H) → **120**; **119** → (R^2 = H) → **121**; **119** → **122** (2.68)

The direct oxidation of phenols (**117a**) to *o*-quinones (**121**) is quite useful. It was used by Harvey in a synthesis of highly carcinogenic polycyclic aromatic diols such as **123** (Eq. 2.69).[68]

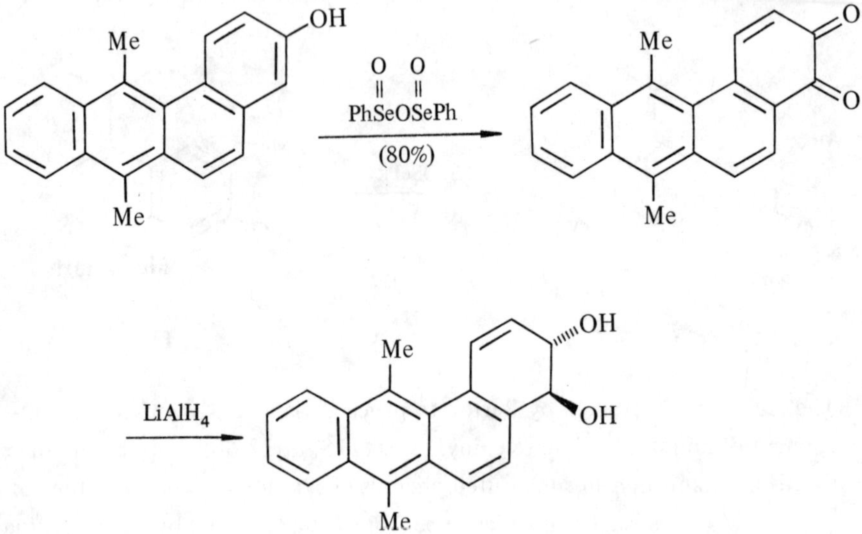

(2.69)

123

In the presence of hexamethyldisilasane, phenols react with benzeneseleninic anhydride to give phenylselenoimines, which can be further reduced to hydroxy anilines (Eq. 2.70).[69]

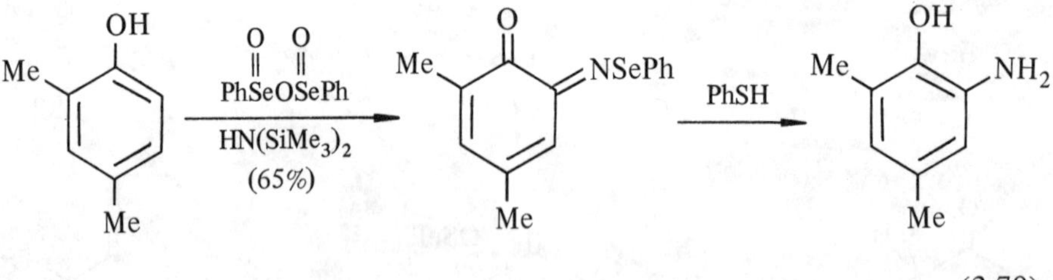

(2.70)

Benzeneseleninic anhydride was also found to effect angular hydroxylation of polycyclic ketones. Thus, treatment of **124** (Eq. 2.71) and **125** (Eq. 2.72) with the reagent did not lead to dehydrogenated products but gave the corresponding α-hydroxy ketones in a rather stereoselective manner.[70] Similarly, chaparrinone triacetate gave **126** (Eq. 2.73).[71]

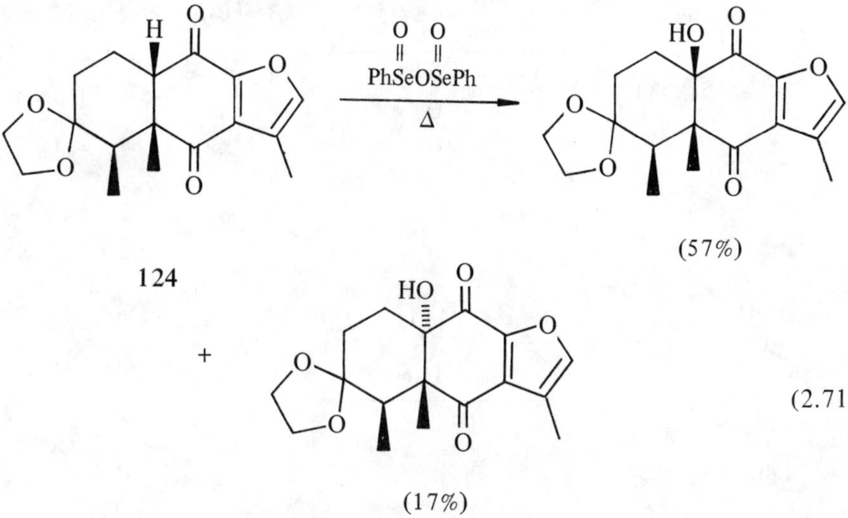

PhSeOSePh (O O) Δ

(57%)

124

+

(17%)

(2.71)

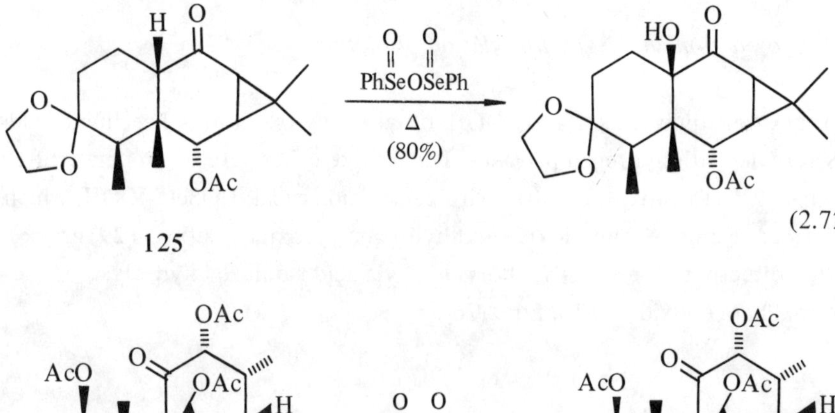

PhSeOSePh (O O) Δ (80%)

OAc

125

(2.72)

PhSeOSePh (O O) (20%)

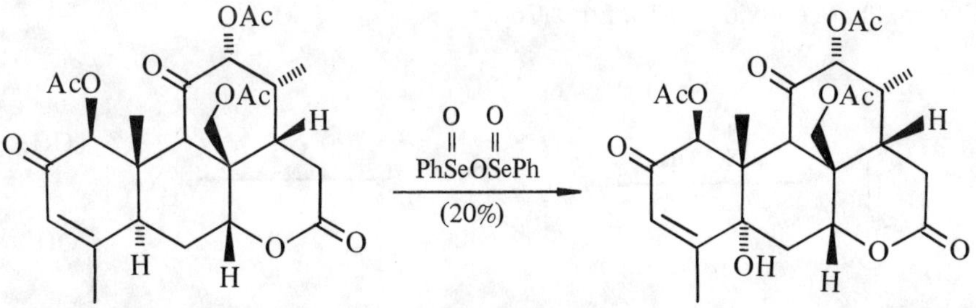

126 (2.73)

Phenols that have more than one hydroxyl group do not follow the pathway shown in Eq. 2.68. Instead, treatment of phenoxides with benzeneseleninic anhydride leads to the formation of either γ-hydroxydiones, such as **127** (Eq. 2.74)[65] or *p*-quinones, such as **128** (Eq. 2.75).[65]

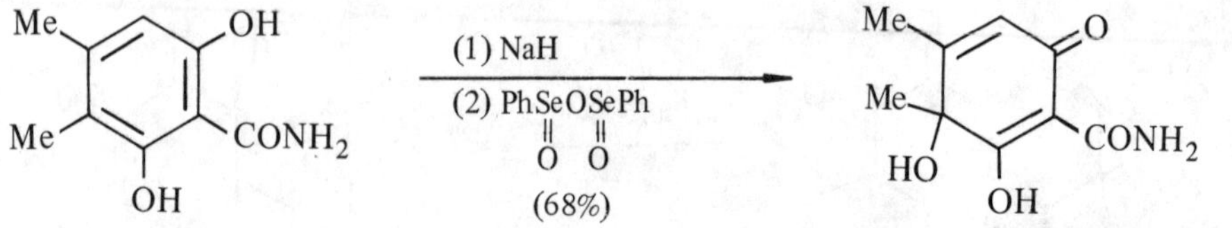

(1) NaH
(2) PhSeOSePh (68%)

127 (2.74)

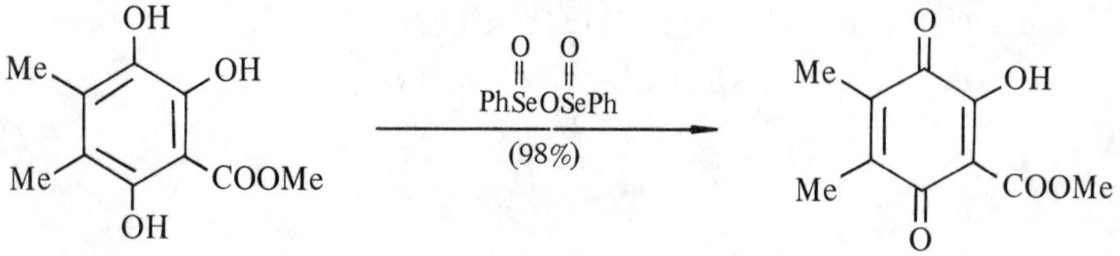

PhSeOSePh (98%)

128 (2.75)

2.5 *Epoxidation and* vic-*hydroxylation of olefins*

Peroxyseleninic acids, RSe(O)OOH, obtained by oxidation of seleninic acids, RSe(O)OH, with hydrogen peroxide, behave like other peracids in converting olefins (**129**) to epoxides (**130**). With perselenious acid, HOSe(O)OOH, which is formed from selenium dioxide and hydrogen peroxide, olefins (**129**) are converted directly to the *trans-vic*-diols (**131**) via acid catalyzed hydrolysis of the intermediate epoxides (**130**, Eq. 2.76).[72]

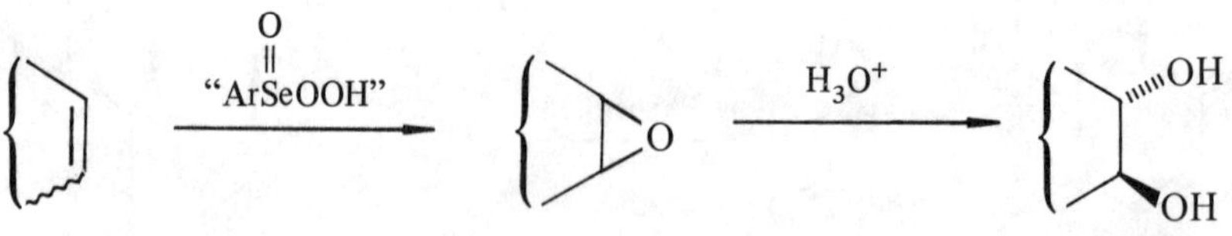

"ArSeOOH" H_3O^+

129 130 131

(2.76)

The vicinal hydroxylation of olefins with selenium dioxide and hydrogen peroxide is not synthetically useful, due to its low efficiency and selectivity. A typical example is the conversion of limonene (**132**) to diol **133** (Eq. 2.77), which is accompanied by several side products.[73]

SeO_2–H_2O_2 ; $H_3O^{\oplus}$

132 133

(2.77)

Treatment of olefins with arylperoxyseleninic acids, however, affords the corresponding epoxides cleanly. Thus, Grieco has developed an epoxidation method based on the use of benzeneperoxyseleninic acid, PhSe(O)OOH, prepared *in situ* by oxidation of phenyl diselenide, PhSeSePh, with hydrogen peroxide. These conditions were quite effective with a variety of olefins, with selectivity comparable to that of other methods. For example, linalool (**134**) gave **135** (Eq. 2.78), while geraniol (**136**) gave the two possible epoxides, **137** and **138**, in 5:1 ratio (Eq. 2.79).[74]

"PhSe(O)OOH" (63%) (2.78)

134 135

"PhSe(O)OOH" (65%)

136 137 (5:1) 138

(2.79)

Sharpless[75] and Reich[76] have shown that the epoxidation of olefins can also be achieved with hydrogen peroxide in the presence of catalytic amounts of arylseleninic acids, ArSe(O)OH. Most effective catalysis was observed with the nitro-substituted ones.[75,76] Kametani has exploited the reaction in an alternative way:[77] the arylseleninic acid required was formed *in situ* by [2,3]-sigmatropic rearrangement of an allylic selenoxide. The latter was obtained by hydrogen peroxide oxidation of the corresponding selenide in the presence of pyridine. The process is illustrated in Eq. 2.80 with the selenide (**139**) derived from geraniol (**136**)[77] by Grieco's procedure.[78]

Se
O_2N
139
H_2O_2–pyr
(96%)
O
Se
O_2N
H_2O_2
(2.80)
OH
O
135
OH
+
O
SeOOH
NO_2

2.6 Baeyer–Villiger type reactions

Another reaction of peroxyseleninic acids, RSe(O)OOH, is the conversion of cyclic ketones (**140**) to lactones (**141**), as in the Baeyer–Villiger reaction.

O
140
O
"RSeOOH"
(R = Ph or OH)
O
O
141
(2.81)

Such a transformation can be effected with hydrogen peroxide and catalytic amounts of selenium dioxide. For example, adamantanone (**142**) was converted to the corresponding lactone (**143**) by using these conditions (Eq. 2.82).[79] Other less strained ketones, however, do not react as cleanly as **142**.

SeO_2 cat–H_2O_2, *t*-BuOH (96%)

142 → **143** (2.82)

Grieco has found that benzeneperoxyseleninic acid, PhSe(O)OOH, formed *in situ* from benzeneseleninic acid, PhSe(O)OH and hydrogen peroxide, is an efficient reagent for Bayer–Villiger type reactions.[80] Thus, estrone methyl ether (**144**) gave lactone **145** in high yield (Eq. 2.83), while ketone **146**, upon prolonged treatment with the reagent at 45° C, formed after esterification with diazomethane, the corresponding hydroxy ester (**147**, Eq. 2.84).[80] Particularly noteworthy is the selective reaction of benzeneperoxyseleninic acid with the carbonyl of **146** without significant interference from the olefinic bond.

"PhSe(O)OOH" (80%)

144 → **145** (2.83)

(1) "PhSe(O)OOH"
(2) CH_2N_2
(63%)

146 → **147** (2.84)

Using this novel reagent, Grieco converted damsin (**148**) to psilostachyin C (**149**, Eq. 2.85).[80,81]

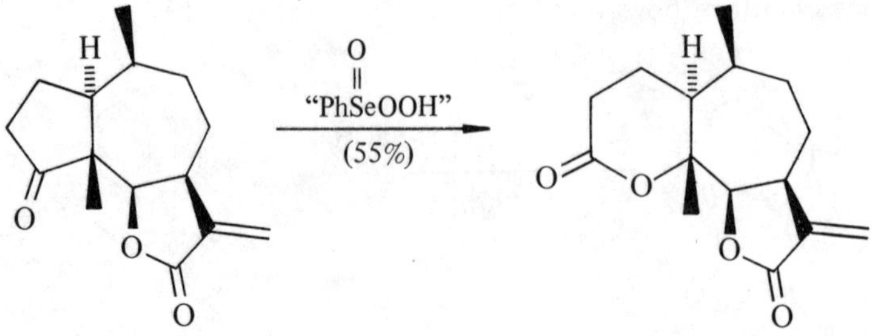

148 **149** (2.85)

Steroidal ketones, such as **150**, react with phenylselenyl chloride, PhSeCl, to give the corresponding α-phenylseleno ketones, which upon oxidation with hydrogen peroxide produce directly the α,β-unsaturated lactones (**151**, Eq. 2.86).[82,83] Apparently, the benzeneperoxyseleninic acid required for the Baeyer–Villiger step is formed *in situ* by oxidation of benzeneseleninic acid, the product of selenoxide *syn*-elimination (Chapter 4).

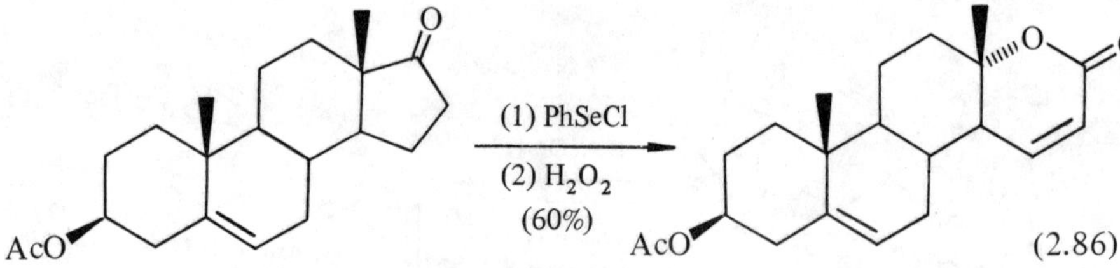

150 **151** (2.86)

REFERENCES

1. (a) Rabjohn, N., "Selenium Dioxide Oxidation" in Dauben, W. G. (ed.), *Organic Reactions*, Wiley, New York (1976), Vol. 24, Chap. 4, p. 261; (b) Jerussi, R. A., in Thyagarajan, B. S. (ed.), *Selective Organic Transformations*, Wiley, New York,(1970), Vol. 1, p. 301; (c) Trachtenberg, E. N., "Selenium Dioxide Oxidation" in Augustine, R. L. (ed.), *Oxidation Techniques and Applications in Organic Synthesis*, Dekker, New York (1969), Vol. 1, Chap.

3, p. 119; (d) Rabjohn, N., "Selenium Dioxide Oxidation" in Adams, R., (ed.), *Organic Reactions*, Wiley, New York (1949), Vol. 5, Chap. 8, p. 331; (e) Waitkins, G. R., and Clark, C. W., *Chem. Rev.,* **36**, 235 (1945).

2. (a) Sharpless, K. B., and Lauer, R. F., *J. Am. Chem. Soc.,* **94**, 7154 (1972); (b) Arigoni, D., Vasella, A., Sharpless, K. B., and Jensen, H. P., *J. Am. Chem. Soc.,* **95**, 7917 (1973).
3. Warpehoski, M. A., Chabaud, B., and Sharpless, K. B., *J. Org. Chem.,* **47**, 2897 (1982).
4. Stephenson, L. M., and Speth, D. R., *J. Org. Chem.,* **44**, 4683 (1979).
5. Sathe, V. M., Chakravarti, K. K., Kadival, M. V., and Bhattacharyya, S. C., *Indian J. Chem.,* **4**, 393 (1966).
6. (a) Plattner, J. J., and Rapoport, H., *J. Am. Chem. Soc.,* **93**, 1758 (1971); (b) Bhalerao, U. T., Plattner, J. J., and Rapoport, H., *J. Am. Chem. Soc.,* **92**, 3429 (1970); (c) Plattner, J. J., Bhalerao, U. T., and Rapoport, H., *J. Am. Chem. Soc.,* **91**, 4933 (1969); (d) Bhalerao, U. T., and Rapoport, H., *J. Am. Chem. Soc.,* **93**, 4835 (1971).
7. Meinwald, J., Opheim, K., and Eisner, T., *Tetrahedron Lett.,* 281 (1973).
8. Miller, G. H., Katzenellenbogen, J. A., and Bowlus, S. B., *Tetrahedron Lett.*, 285 (1973).
9. Mori, K., Ohki, M., and Matsui, M., *Tetrahedron,* **30**, 715 (1974).
10. Knight, D. W., and Rustidge, D. C., *J. C. S. Perkin I*, 679 (1981).
11. Corey, E. J., Tius, M. A., and Das, J., *J. Am. Chem. Soc.,* **102**, 1742 (1980).
12. Corey, E. J., Tius, M. A., and Das, J., *J. Am. Chem. Soc.,* **102**, 7612 (1980).
13. Clark, K. J., Fray, G. I., Jaeger, R. H., and Robinson, R., *Tetrahedron,* **6**, 217 (1959).
14. Thomas, A. F., *J. C. S. Chem. Comm.,* 947 (1967).
15. Corey, E. J., and Snider, B. B., *J. Am. Chem. Soc.,* **94**, 2549 (1972).
16. Elliott, M., Janes, N. F., and Pullman, D. A., *J. C. S. Perkin I,* 2470 (1974).
17. Bhalerao, U. T., and Rapoport, H., *J. Am. Chem. Soc.,* **93**, 5311 (1971).
18. Corey, E. J., Gilman, N. W., and Ganem, B. E., *J. Am. Chem. Soc.,* **90**, 5616 (1968).
19. Toth, J. O., Luu, B., and Ourisson, G., *Tetrahedron Lett.,* **24**, 1081 (1983).
20. Smith, C. W., and Holm, R. T., *J. Org. Chem.,* **22**, 747 (1957).
21. Kazlauskas, R., Pinhey, J. T., Simes, J. J. H., and Watson, T. G., *J. C. S. Chem. Comm.,* 945 (1949).
22. Schulte, K. H., Gatola, M., and Ohloff, G., *Helv. Chim. Acta,* **56**, 2028 (1973).

23. Büchi, G., and West, H., *J. Org. Chem.*, **34**, 857 (1969).
24. Takayanagi, H., and Nishino, C., *J. Chem. Ecol.*, **8**, 883 (1982).
25. Kishi, Y., Aratoni, M., Fukuyama, T., Nakatsubo, F., Goto, T., Inoue, S., Tanino, H., Sugiura, S., and Kakoi, H., *J. Am. Chem. Soc.*, **94**, 9217, 9219 (1972).
26. Abe, Y., Harukawa, T., Ishikawa, H., Miki, T., Sumi, M., and Toga, T., *J. Am. Chem. Soc.*, **78**, 1422 (1956).
27. Doyle, P., Maclean, J. R., Parker, W., and Raphael, R. A., *Proc. Chem. Soc.*, 239 (1963).
28. Danieli, N., Mazur, Y., and Sondheimer, F., *J. Am. Chem. Soc.*, **84**, 875 (1962).
29. Danieli, N., Mazur, Y., and Sondheimer, F., *Tetrahedron Lett.*, 310 (1961).
30. Tankard, M. H., and Whitehurst, J. S., *Tetrahedron*, **30**, 451 (1974).
31. Evans, D. A., and Sims, C. L., *Tetrahedron Lett.*, 4691 (1973).
32. Martin, S. F., Desai, R. S., Philips, G. W., and Miller, A. C., *J. Am. Chem. Soc.*, **102**, 3294 (1980).
33. Tang, C., and Rapoport, H., *J. Am. Chem. Soc.*, **94**, 8615 (1972).
34. Umbreit, M. A., and Sharpless, K. B., *J. Am. Chem. Soc.*, **99**, 5526 (1977).
35. Coxon, J. M., and Dansted, E., *Org. Syn.*, **56**, 25 (1976).
36. Snider, B. B., and Duncia, J. V., *J. Org. Chem.*, **45**, 3461 (1980).
37. Haruna, M., and Ho, K., *J. C. S. Chem. Comm.*, 483 (1981).
38. Shirahama, H., Hayano, K., Arora, G. S., Ohtsuka, T., Murata, Y., and Matsumoto, T., *Chem. Lett.*, 1417 (1982).
39. Paaren, H. E., DeLuca, H. F., and Schnoes, H. K., *J. Org. Chem.*, **45**, 3253 (1980).
40. Sum, F. W., and Weiler, L., *J. Am. Chem. Soc.*, **101**, 4401 (1979).
41. Greenlee, M. L., *J. Am. Chem. Soc.*, **103**, 2425 (1981).
42. Tanaka, K., Uchiyama, F., Sakamoto, K., and Inubushi, Y., *J. Am. Chem. Soc.*, **104**, 4965 (1982).
43. Matsumoto, T., Imai, S., and Yuki, S., *Bull. Chem. Soc. Japan*, **54**, 1448 (1981).
44. Kumonaka, T., Kanai, Y., Yanagiya, M., and Matsumoto, T., *Chem. Lett.*, 1715 (1982).
45. Chhabra, B. R., Hayano, K., Ohtsuka, T., Shirahama, H., and Matsumoto, T., *Chem. Lett.*, 1703 (1981).
46. Chabaud, B., and Sharpless, K. B., *J. Org. Chem.*, **44**, 4202 (1979).
47. Clar, E., and Stewart, D. G., *J. Chem. Soc.*, 687 (1951).

48. Burger, A., and Madlin, L. R., Jr., *J. Am. Chem. Soc.*, **62**, 1079 (1940).
49. Jerchel, D., Heider, J., and Wagner, M. V., *Ann. Chem.*, **613**, 153 (1958).
50. Forbis, R. M., and Rinehart, K. C., Jr., *J. Am. Chem. Soc.*, **95**, 5003 (1973).
51. (a) Cain, M., Campos, O., Guzman, F., and Cook, J. M., *J. Am. Chem. Soc.*, **105**, 907 (1983); (b) Campos, O., and Cook, J. M., *Tetrahedron Lett.*, 1025 (1979).
52. (a) Büchi, G., Foulkes, D. M., Kurono, M., Mitchell, G. F., and Schneider, R. S., *J. Am. Chem. Soc.*, **89**, 6745 (1967); (b) Büchi, G., Foulkes, D. M., Kurono, M., Mitchell, G. F., *J. Am. Chem. Soc.*, **88**, 4534 (1966).
53. Suzuki, E., Hamajima, K., and Inoue, S., *Synthesis*, 192 (1975).
54. (a) Barton, D. H. R., Hui, R. A. H. F., and Ley, S. V., *J. C. S. Perkin I*, 2179 (1982); (b) Barton, D. H. R., Hui, R. A. H. F., Lester, D. I., and Ley, S. V., *Tetrahedron Lett.*, 3331 (1979).
55. Corey, E. J., and Schaefer, J. P., *J. Am. Chem. Soc.*, **82**, 918 (1960).
56. Sharpless, K. B., and Gordon, K. M., *J. Am. Chem. Soc.*, **98**, 300 (1976).
57. Marguet, A., Dvolaitzky, M., and Arigoni, D., *Bull. Soc. Chim. France*, 2956 (1966).
58. Allen, M. S., Lamb, N., Money, T., and Salisbory, P., *J. C. S. Chem. Comm.*, 112 (1979).
59. Polonski, T., *J. C. S. Perkin I*, 305 (1983).
60. Ayer, W. A., Bowman, W. R., Joseph, T. C., and Smith, P., *J. Am. Chem. Soc.*, **90**, 1648 (1968).
61. Woodward, R. B., Cava, M. P., Ollis, W. D., Hunger, A., Daeniker, H. U., and Schenker, K., *Tetrahedron*, **19**, 247 (1963); (b) Woodward, R. B., Cava, M. P., Ollis, W. D., Hunger, A., Daeniker, H. U., and Schenker, K., *J. Am. Chem. Soc.*, **76**, 4749 (1954).
62. Hallmann, G., and Hagele, K., *Ann. Chem.*, **662**, 147 (1963).
63. Howe, R., and Johnson, D., *J. C. S. Perkin I*, 977 (1972).
64. Buu-Hoi, N. P., Saint-Ruf, G., and Bourgeade, J. C., *J. Heterocyclic Chem.*, **5**, 545 (1968).
65. (a) Barton, D. H. R., Ley, S. V., Magnus, P. D., and Rosenfeld, M. N., *J. C. S. Perkin I*, 567 (1977); (b) Barton, D. H. R., Magnus, P. D., and Rosenfeld, M. N., *J. C. S. Chem. Comm.*, 301 (1975).
66. Jeffs, P. W., and Lynn, D. G., *Tetrahedron Lett.*, 1617 (1978).
67. (a) Barton, D. H. R., Brewster, A. G., Ley, S. V., Read, C. M., and Rosenfeld, M. N., *J. C. S. Perkin I*, 1473 (1981); (b) Barton, D. H. R., Brewster, A. G., Ley, S. V., and Rosenfeld, M. N., *J. C. S. Chem. Comm.*, 985 (1976).

68. Sukumaran, K. B., and Harvey, R. G., *J. Org. Chem.,* **45**, 4407 (1980); *J. Am. Chem. Soc.,* **101**, 1353 (1979).
69. (a) Barton, D. H. R., Brewster, A. G., Ley, S. V., and Rosenfeld, M. N., *J. C. S. Chem. Comm.,* 147 (1977); (b) Holker, J. S. E., O'Brien, E., and Park, B. K., *J. C. S. Perkin I,* 1915 (1982).
70. (a) Yamakawa, K., Satoh, T., Ohba, N., Sakaguchi, R., Takita, T., and Tamura, N., *Tetrahedron,* **37**, 473 (1981); (b) Yamakawa, K., Satoh, T., Ohba, N., and Sakaguchi, R., *Chem. Lett.,* 763 (1979).
71. Khoi, N., and Polonsky, J., *Helv. Chim. Acta,* **64**, 1540 (1981).
72. (a) Hakura, J., Tanaka, H., and Ito, H., *Bull. Chem. Soc. Japan,* **42**, 1604 (1969); (b) Sonoda, N., and Tsutsumi, S., *Bull. Chem. Soc. Japan,* **38**, 958 (1965); (c) Stoll, A., Lindermann, A., and Jucker, E., *Helv. Chim. Acta,* **36**, 268 (1953); (d) Mugdan, M., and Young, D. P., *J. Chem. Soc.,* 2988 (1949).
73. (a) Wilson, C. W., III, and Shaw, P. E., *J. Org. Chem.,* **38**, 1684 (1973); (b) Sumimoto, M., Suzuki, T., and Kondo, T., *Agr. Biol. Chem.,* **38**, 1061 (1974).
74. Grieco, P. A., Yokoyama, Y., Gilman, S., and Nishizawa, M., *J. Org. Chem.,* **42**, 2034 (1977).
75. Hori, T., and Sharpless, K. B., *J. Org. Chem.,* **43**, 1689 (1978).
76. (a) Reich, H. J., Wollowitz, S., Trend, J. E., Chow, F., and Wendelborn, D. F., *J. Org. Chem.,* 1697 (1978); (b) Reich, H. J., Chow, F., and Peake, S. L., *Synthesis,* 299 (1978).
77. Kametani, T., Nemoto, H., and Fukumoto, K., *Bioorganic Chem.,* **7**, 215 (1978); *Heterocycles,* **6**, 1365 (1977).
78. Grieco, P. A., Gilman, S., and Nishizawa, M., *J. Org. Chem.,* **41**, 1485 (1976). See also § 4.4.
79. Faulkner, D., and McKervey, M. A., *J. C. S. Chem. Comm.,* 3906 (1971).
80. Grieco, P. A., Yokoyama, Y., Gilman, S., and Ohfune, Y., *J. C. S. Chem. Comm.,* 870 (1977).
81. Grieco, P. A., Ohfune, Y., and Majetich, G., *J. A. C. S.,* **99**, 7393 (1977).
82. Sharpless, K. B., Gordon, K. M., Lauer, R. F., Patrick, D. W., Singer, S. P., and Young, M. W., *Chem. Scr.,* **8A**, 9 (1975).
83. Williams, J. R., and Leber, J. D., *Synthesis,* 427 (1977).
84 (a) Schmuff, N. R., and Trost, B. M., *J. Org. Chem.,* **48**, 1404 (1983); (b) Trost, B. M., and Verhoeven, T. R., *J. Am. Chem. Soc.,* **100**, 3435 (1978).
85. (a) Wittek, P. J., Liao, T. K., and Cheng, C. C., *J. Org. Chem.,* **44**, 870 (1979); *J. Heterocycl. Chem.,* **13**, 1283 (1976); (b) Kende, A. S., Lorah, D. P., and Boatman, R. J., *J. Am. Chem. Soc.,* **103**, 1271 (1981).

CHAPTER 3

Selenium-mediated Dehydrogenations

Selenium dioxide, SeO_2, and benzeneseleninic anhydride, PhSe(O)OSe(O)Ph, are not only useful for the introduction of oxygen into organic molecules (Chapter 2), but they are also useful for introducing unsaturation into certain substrates by an overall removal of two hydrogen atoms. Another selenium reagent capable of effecting this transformation is diphenylselenium bis(trifluoroacetate), $Ph_2Se(OCOCF_3)_2$, recently introduced by Marino.[31] Dehydrogenation reactions using the above reagents are described in this chapter. The alternative means of introducing an olefinic bond, by the stepwise attachment and oxidative removal of an organoselenium functionality, are discussed in Chapter 4.

3.1 *Dehydrogenation of carbonyl compounds*

A common side reaction of the *a*-oxygenation of ketones (**1**) with selenium dioxide[1] is dehydrogenation to the corresponding enone (**4**). The mechanism of the reaction was examined by several investigators.[2, 3] In the most recent report by Sharpless,[3] it was formulated as an electrophilic attack of selenium dioxide (or its solvated derivative) to the ketone (**1**) or its enol form (**2**), followed by *syn*-elimination of the resulting ketoseleninic acid (**3**), as shown below in Eq. 3.1. With certain sterically constrained substrates, such as steroidal ketones, this is usually the main pathway of the reaction. Also, the de-

1 ⇌ 2 → 3 —$-Se(OH)_2$→ 4

(3.1)

hydrogenated product is favored if the oxidation is carried out in *t*-butanol as solvent.[4] Steroidal ketones of type 5, or enones of type 7, can be dehydrogenated under these conditions to the corresponding steroidal dienones (6) in one step and in good yield (Eq. 3.2).[4]

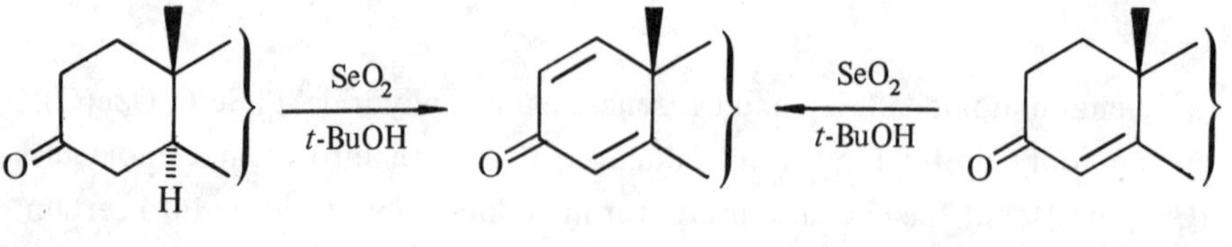

SeO2
t-BuOH
H
O
SeO2
t-BuOH
O
O

5 6 7

(3.2)

Dehydrogenations of this type were used recently by Smith in the synthesis of (±)-mayurone (8, Eq. 3.3)[5] and several pentenomycin antibiotics such as (±)-pentenomycin I (9, Eq. 3.4).[6]

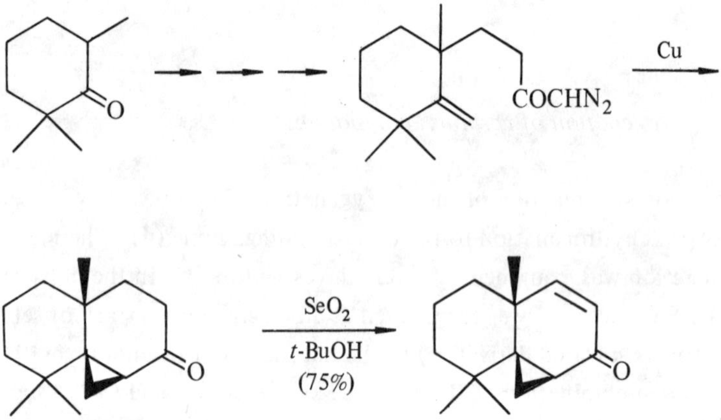

O
COCHN2
Cu
SeO2
t-BuOH
(75%)
O
O

(3.3)

8

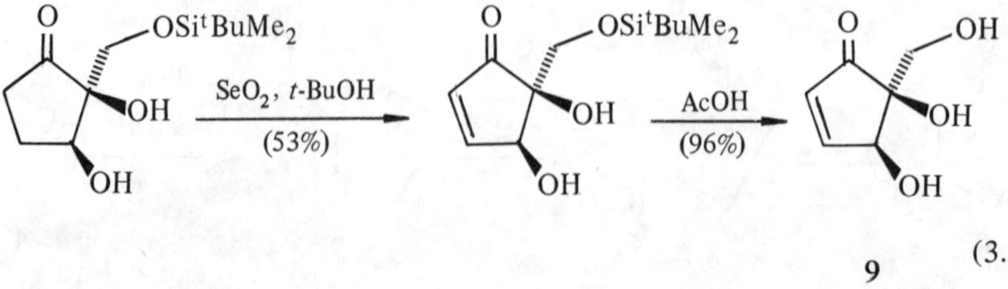

O
OSitBuMe2
OH
OH
SeO2, t-BuOH
(53%)
O
OSitBuMe2
OH
OH
AcOH
(96%)
O
OH
OH
OH

9

(3.4)

Other reactions of selenium dioxide can sometimes take place in addition to the dehydrogenation step. For example, oxidation of **10** gives the *α*-dione, which was further dehydrogenated to furnish mansonone D (**11**, Eq. 3.5).[7] Similarly, Kishi converted **12** to **13**, which is a precursor of the aromatic segment (**14**, Eq. 3.6) required in his rifamycin synthesis.[8]

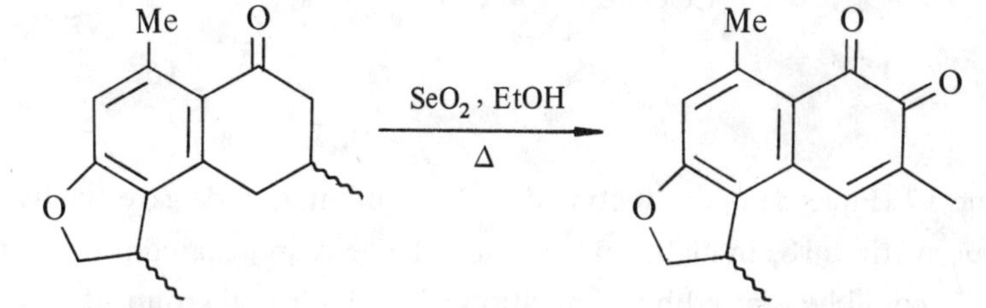

SeO_2, EtOH, Δ

10 → **11** (3.5)

SeO_2, AcOH (53%)

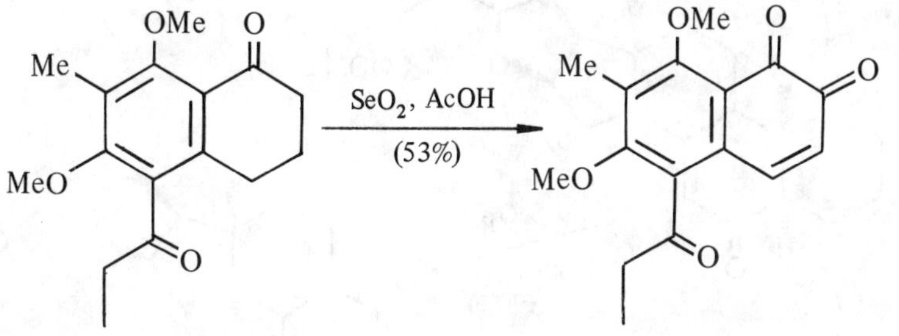

12 → **13**

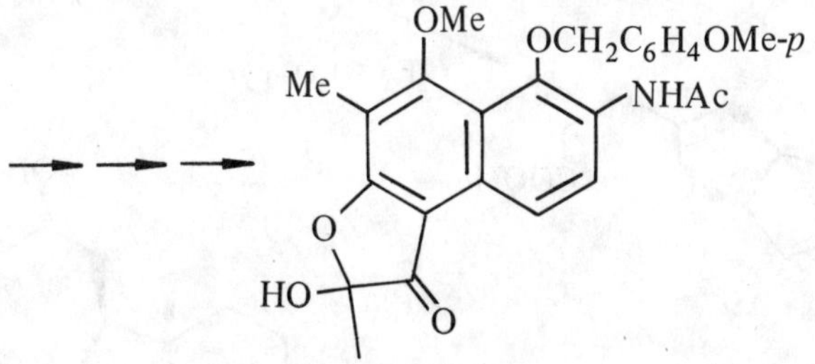

→ → → **14** (3.6)

Allylic oxidation of **15** with allylic rearrangement, followed by dehydrogenation, gave after lactonization, intermediate **16** (Eq. 3.7),[9] which was converted to camptothecin (**64**, Eq. 2.29).

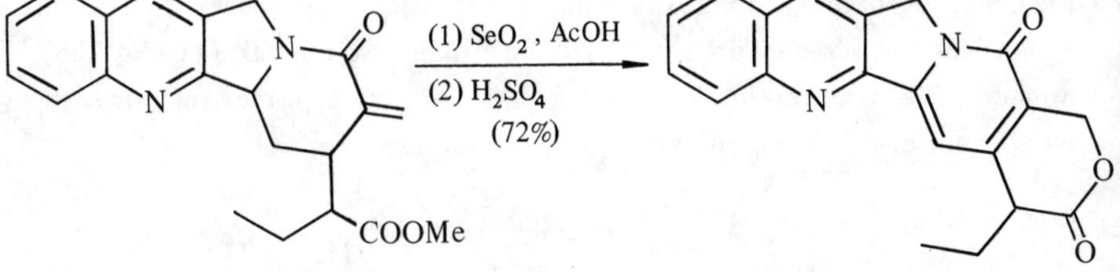
N
O
COOMe
(1) SeO_2, AcOH
(2) H_2SO_4
(72%)
N
O
O
O

15 **16**

(3.7)

Enone **17** (Eq. 3.8) upon treatment with selenium dioxide gave the hydroxylated derivative (**18**) in high yield, and not the dehydrogenated product (**19**).[10] The latter could be formed by elimination of the hydroxyl group; it was used to prepare α-ecdysone (**20**, Eq. 3.8).[10]

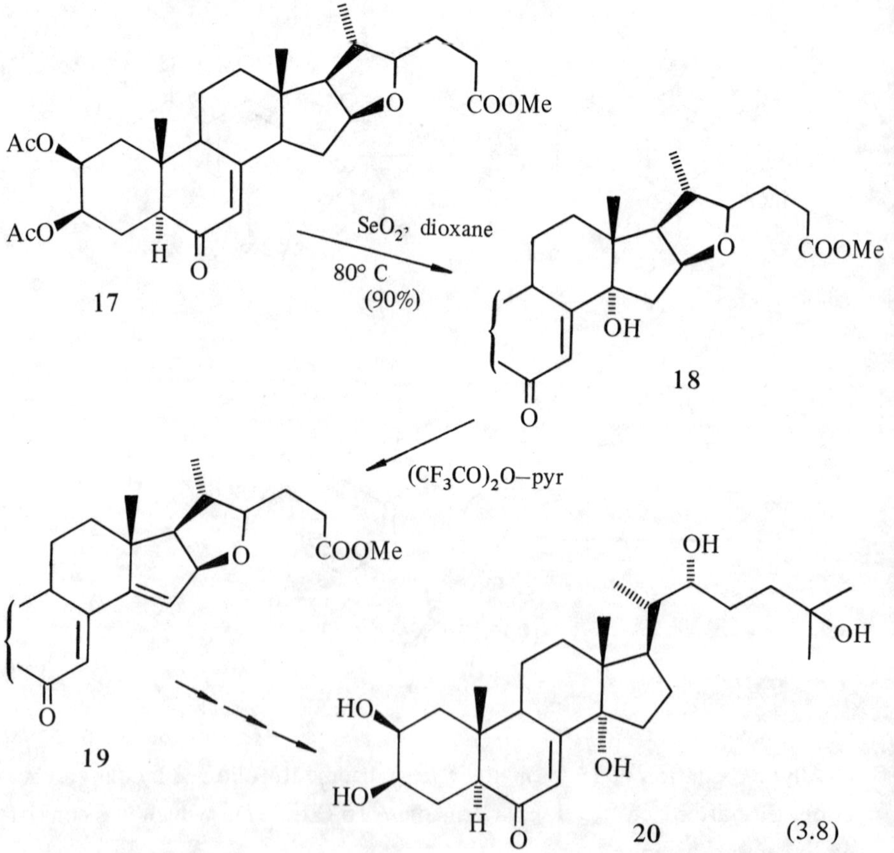
O
COOMe
AcO
AcO
H
O
SeO_2, dioxane
80° C
(90%)
O
COOMe
OH
O
$(CF_3CO)_2O$–pyr
O
COOMe
O
OH
OH
HO
HO
OH
H
O

17 **18** **19** **20**

(3.8)

The presence of a second carbonyl group in a position that would be in conjugation with the new olefinic bond, makes the dehydrogenation reaction a highly favored process. Thus, 1,4-dicarbonyl systems of type **21** are readily dehydrogenated with selenium dioxide to give the corresponding conjugated products (**22**, Eq. 3.9). This type of transformation was used by Smith in the synthesis of the novel antibiotic mycorrhizin A (**26**, Eq. 3.10).[11] Treatment of intermediate **23** with selenium dioxide in *t*-butanol–pyridine introduced the desired olefinic bond, but it also partially hydrolyzed the mixed methyl ketal to give a 2:1 mixture of **24** and **25**. Both constitutents were converted to **26** (Eq. 3.10).[11]

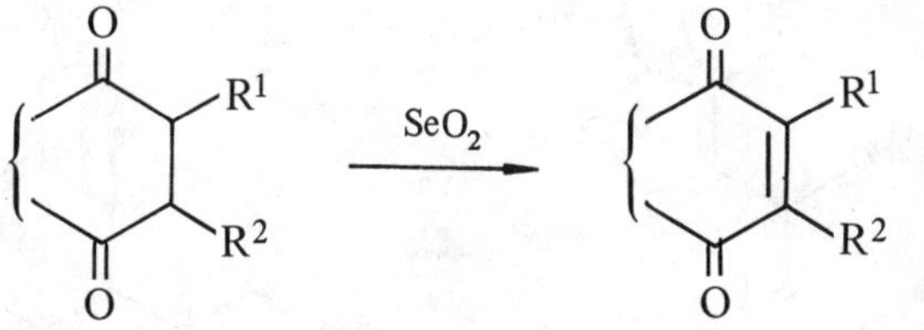

SeO_2

(3.9)

21 **22**

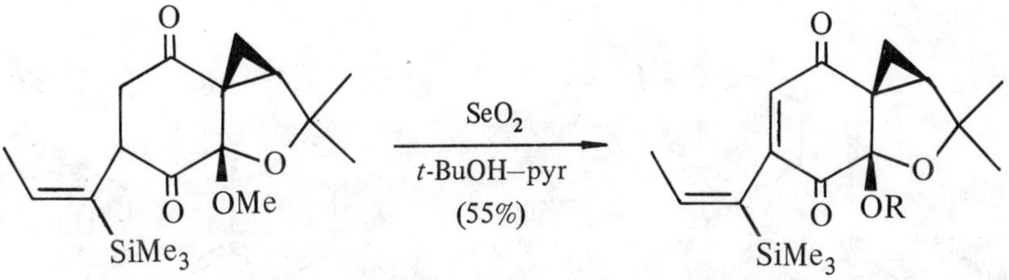

SeO_2
t-BuOH–pyr
(55%)

23

24: R = Me
25: R = H

(1) Cl_2
(2) KF

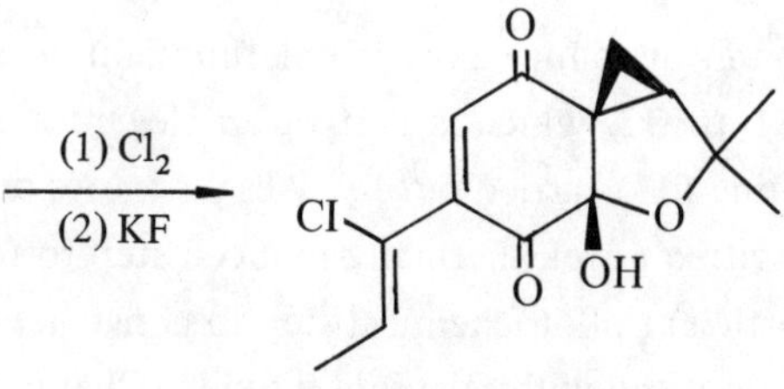

(3.10)

26

A stereochemical requirement for the formation of compounds of type 22, in which both R^1 and R^2 are non-hydrogen substituents, is that R^1 and R^2 should be *cis* in the precursor (21), otherwise the reaction may take a different course. For example, steroidal diones of type 27 and 28 give different types of products on treatment with selenium dioxide, as shown in Eq. 3.11 and Eq. 3.12, respectively.[12]

The correct arrangement (*cis*) of the two hydrogens in 27 allows the formation of the fully conjugated dione (Eq. 3.11), while the *trans* relationship of these hydrogens in 28 leads to the non-fully conjugated enone (Eq. 3.12).

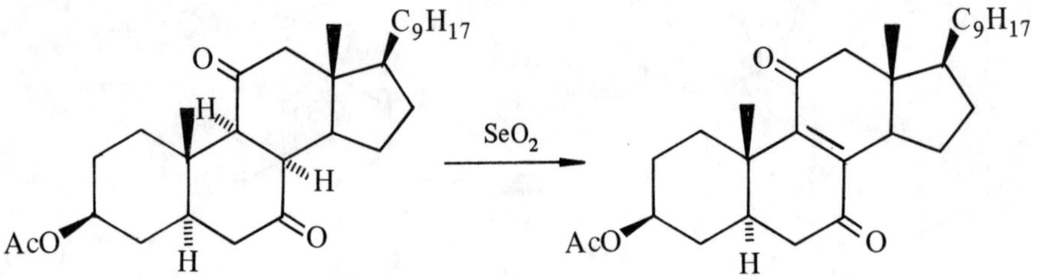

C_9H_{17} O H AcO H O 27 SeO_2 C_9H_{17} O AcO H O (3.11)

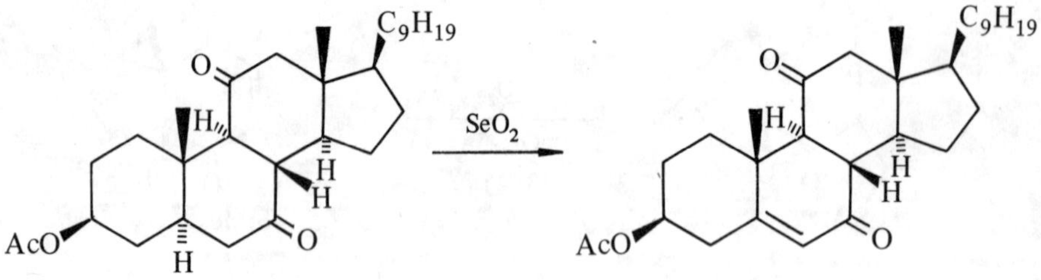

C_9H_{19} O H AcO H H H O 28 SeO_2 C_9H_{19} O H AcO H H O (3.12)

Further dehydrogenation to introduce a second olefinic bond is also possible, as in the conversion of 31 to 32, which is part of an elegant synthesis of the lycopodium alkaloid annotine (33) carried out by Wiesner.[13] As outlined in Eq. 3.13, the synthesis begins with a novel, thermally induced step to form the lactam ring, followed by an efficient photochemical step to construct the cyclobutane ring. An allylic oxidation step with selenium dioxide (Chapter 2) is also utilized for the conversion of 29 to enone 30.

(1) $CH_2{=}CHCOOH$, Δ
(2) $CH_2{=}C{=}CH_2$, *hv*
(100%)

29

(1) SeO_2, AcOH
(2) KOH
(3) $CrO_3 \cdot 2pyr$

30

(1) KCN
(2) H_2SO_4–MeOH

31

SeO_2

32

33

(3.13)

2-Substituted 1,3-dicarbonyl systems can be dehydrogenated with selenium dioxide in a similar manner, giving access to useful synthetic intermediates such as **34** (Eq. 3.14).[14]

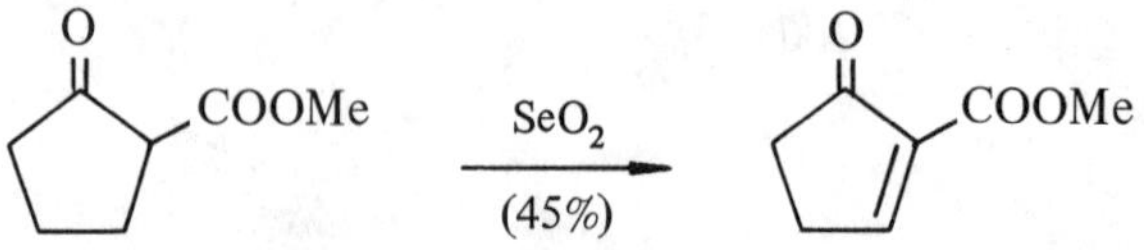

(3.14)

A good alternative to selenium dioxide for the dehydrogenation of ketones is benzeneseleninic anhydride, PhSe(O)OSe(O)Ph, recently used by Barton.[15] The reagent has the advantage that it does not affect olefins and gives the desired enones or dienones in high yield. For example, cholest-4-ene-3-one (**35**) gives dienone **36** in 92% yield (Eq. 3.15).[15]

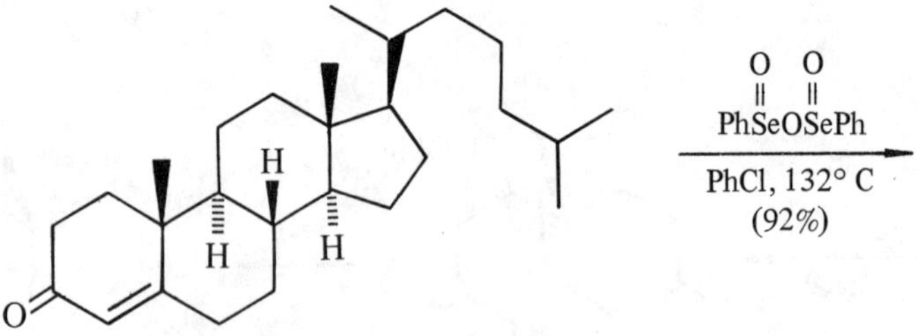

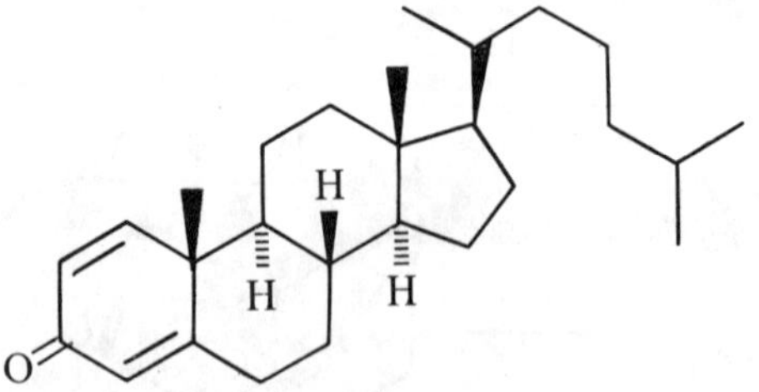

(3.15)

It has been proposed that the reaction proceeds via C-selenenylation of the enolic form (**38**) to give selenoxide **39** , which readily undergoes *syn*-elimination (Chapter 4) to give enone **40** (Eq. 3.14).[15]

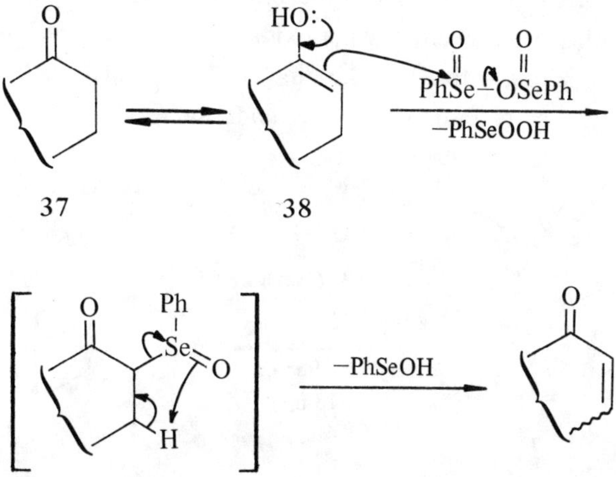

HO:

O O
PhSe–OSePh
–PhSeOOH

37 38

Ph
Se
O
H
–PhSeOH

39 40

(3.16)

If the ketone is α-substituted (as in **41**, R = alkyl), then the oxygen of the enol (**42**) reacts with the anhydride (O-selenenylation) forming **43**, which leads to the α-hydroxy ketone (**44**, Eq. 3.17) via a [2,3]-sigmatropic rearrangement. and cleavage of the resulting oxygen–selenium bond. The overall conversion of **41** to **44** (angular hydroxylation) was indeed observed (Eq. 2.71 and Eq. 2.72).

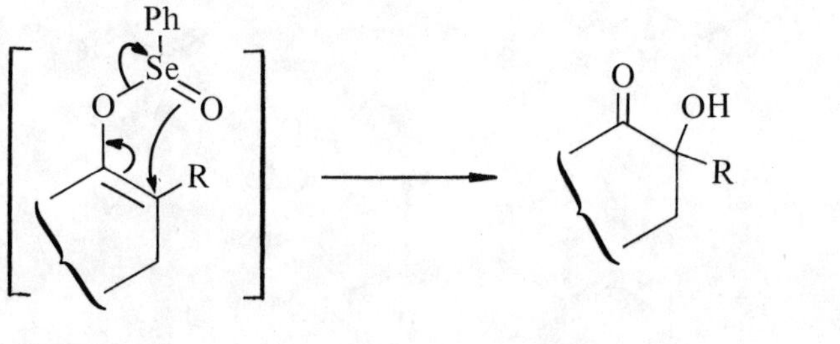

HO:
R R
O O
PhSe–OSePh
–PhSeOOH

41 42

Ph
Se
O O
R
OH
R

43 44

(3.17)

With certain substrates and under certain conditions, particularly excess anhydride and longer reaction times, the initially formed enone (**40**) is transformed into other products by further oxidation. For example, β-amyrone (**45**) forms enone **46**, along with A-nor-2,3-dione (**47**), of which the latter results from a benzylic-type reaction of the overoxidation product of **46** (Eq. 3.18).[15]

(3.18)

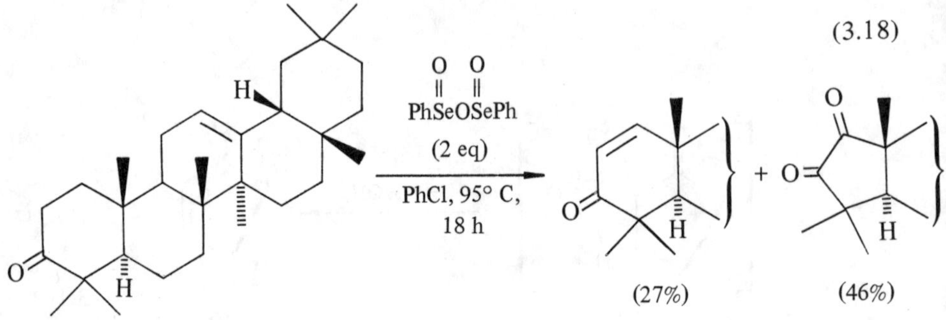

PhSe(O)OSe(O)Ph (2 eq), PhCl, 95° C, 18 h

45 → **46** (27%) + **47** (46%)

Barton and co-workers have also shown that the benzeneseleninic anhydride required for dehydration can be formed *in situ* by dehydration of benzeneseleninic acid or by oxidation of catalytic amounts of diphenyl diselenide with excess *t*-butyl hydroperoxide.[15] More recently, the same investigators have reported the use of iodoxybenzene, $PhIO_2$, and *m*-iodoxybenzoic acid, *m*-O_2I-C_6H_4COOH as catalysts in the direct conversion of ketones or enones to dienones (Eqs. 3.19, 3.20 and 3.21).[16]

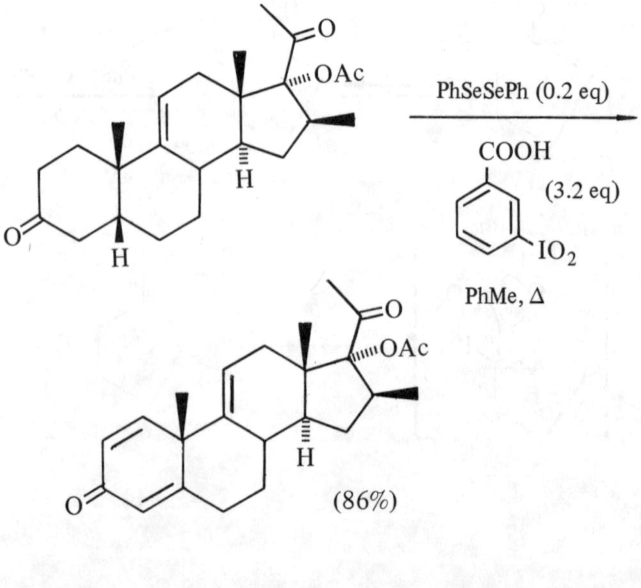

PhSePh (0.2 eq), *m*-$IO_2C_6H_4COOH$ (3.2 eq), PhMe, Δ

(86%)

(3.19)

$$\xrightarrow[\text{PhIO}_2\ (3.6\ \text{eq}),\ \text{benzene},\ \Delta]{\text{PhSe(O)OSe(O)Ph}\ (0.16\ \text{eq})}$$

(72%) (3.20)

$$\xrightarrow[\text{m-IO}_2\text{C}_6\text{H}_4\text{COOH},\ \text{PhMe},\ \Delta]{\text{PhSe(O)OSe(O)Ph}\ (0.1\ \text{eq})}$$

(70%) (3.21)

δ-Lactones are efficiently dehydrogenated with benzeneseleninic anhydride, as shown in Eq. 3.22.[17] γ-Lactones or carboxylic esters, however, are unreactive under these conditions.[17]

$$\xrightarrow[\text{PhCl},\ 100^\circ\ \text{C}]{\text{PhSe(O)OSe(O)Ph}}$$

(90%) (3.22)

Lactams are dehydrogenated similarly to the lactones, as indicated in Eq. 3.23.[18]

$$\xrightarrow[\text{diglyme},\ 120^\circ\ \text{C}]{\text{PhSe(O)OSe(O)Ph}}$$

(56–88%) (3.23)

3.2 Oxidation of alcohols

Benzeneseleninic anhydride oxidizes alcohols (**48**) to the corresponding carbonyl compounds (**50**) via the initially formed seleninic esters (**49**), as shown in Eq. 3.24.[19]

O O
|| ||
PhSeOSePh

OH H SePh O H O O

(3.24)

48 **49** **50**

For example, **51** is oxidized to **52** in high yield, without any reaction at the lactam site, as indicated in Eq. 3.25.[19,18a]

OH

O O
|| ||
PhSeOSePh (1.1 eq)
THF, Δ
(94%)

O N H O N H O

(3.25)

51 **52**

Ketones can undergo further dehydrogenation reactions if they are appropriately substituted. Thus, **53** gives directly the enone (**54**) in good yield, without any interference from the olefin (Eq. 3.26).[19]

C_8H_{17} C_8H_{17}

O O
|| ||
PhSeOSePh
(60%)

HO O

(3.26)

53 **54**

The catalytic system empolyed by Barton (PhSeSePh–$ArIO_2$) is even more effective for the direct conversion of alcohols to enones. A rather spectacular example is shown in Eq. 3.27, where diol **55** is transformed into triene-dione **56** in 64% yield,[16] representing an average yield of *ca.* 97% for each of the 13 intermediate steps! It is also noteworthy that the carboxylic ester functionality did not interfere with the reaction.

OH COOMe PhSeSePh (0.1 eq) m-$O_2IC_6H_4COOH$ (2.5 eq), PhMe, Δ HO H **55** O COOMe O **56** (3.27)

An alternative selenium-based method for the mild oxidation of alcohols was reported recently by Kuwajima involving the use of bis(2,4,6-trimethylphenyl) diselenide (0.5 eq) and a slight excess of *t*-butyl hydroperoxide.[20] Using this method, geraniol (**57**) is quantitatively oxidized to geranial (**58**), while diol **59** is selectively oxidized at the allylic alcohol site (Eq. 3.29).[20] A major advantage of the method is that it allows selective oxidation of alcohols in the presence of a phenylthio or a phenylseleno group. An example is given in Eq. 3.30.[20] Recently a modification of the Corey–Kim oxidation was reported, involving the use of the dimethyl selenide–*N*-chlorosuccinimide complex, which was also found to be a mild oxidant for alcohols.[33]

OH $(2,4,6\text{-}Me_3C_6H_2Se)_2$ (0.5 eq) *t*-BuOOH (1.1 eq), benzene, Δ (100%) CHO **57** **58** (3.28)

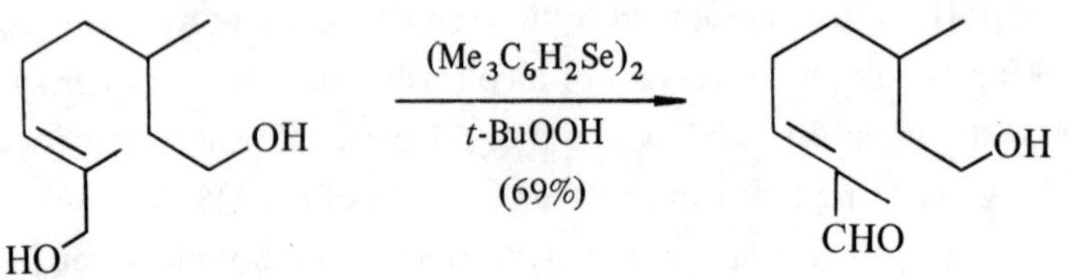

59

$$\xrightarrow[\text{(69\%)}]{(Me_3C_6H_2Se)_2,\ t\text{-BuOOH}}$$ (3.29)

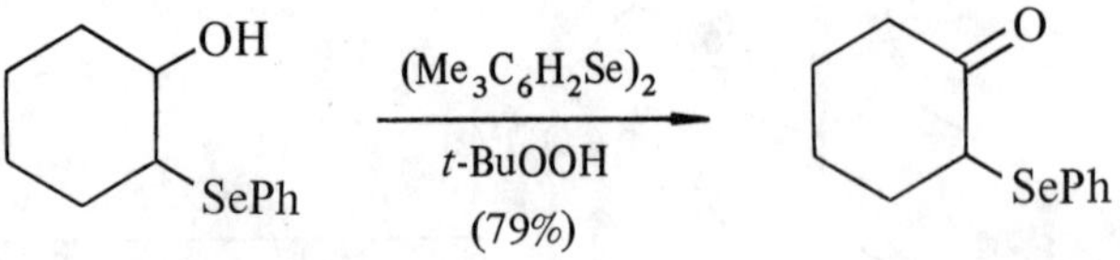

OH, HO, OH, CHO

$$\xrightarrow[\text{(79\%)}]{(Me_3C_6H_2Se)_2,\ t\text{-BuOOH}}$$ (3.30)

OH, SePh, O, SePh

3.3 Cyclization–dehydrogenation of phenolic enones

Phenolic enones of type **60** are converted to the corresponding chromones (**61**, Eq. 3.31) by treatment with selenium dioxide. The reaction is useful for the direct conversion of chalcones to the corresponding flavones.[21] For example, genkwanin (**63**) was synthesized from **62** as outlined in Eq. 3.32.[22] Two more recent examples are given in Eq. 3.33[23] and Eq. 3.34.[24]

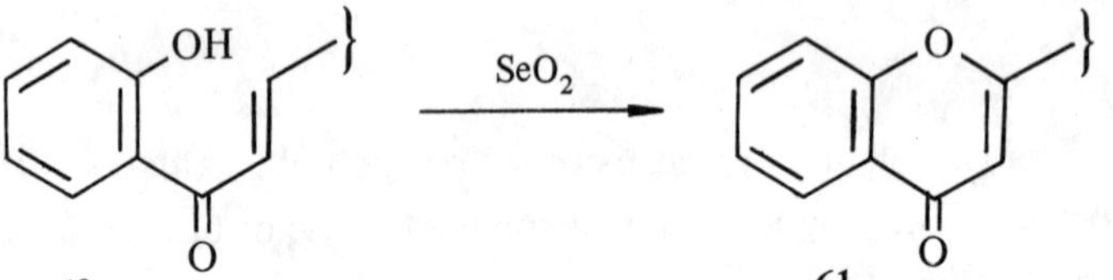

$$\xrightarrow{SeO_2}$$ (3.31)

OH, O, O, O

60 61

$$\xrightarrow[\text{(72\%)}]{SeO_2}$$

MeO, OH, OMe, O, OCH_2Ph, 62, MeO, O, OMe, O, OCH_2Ph

$$\xrightarrow[(2)\ AlCl_3]{(1)\ HCl,\ AcOH}$$ (3.32)

MeO, O, OH, O, OH, 63

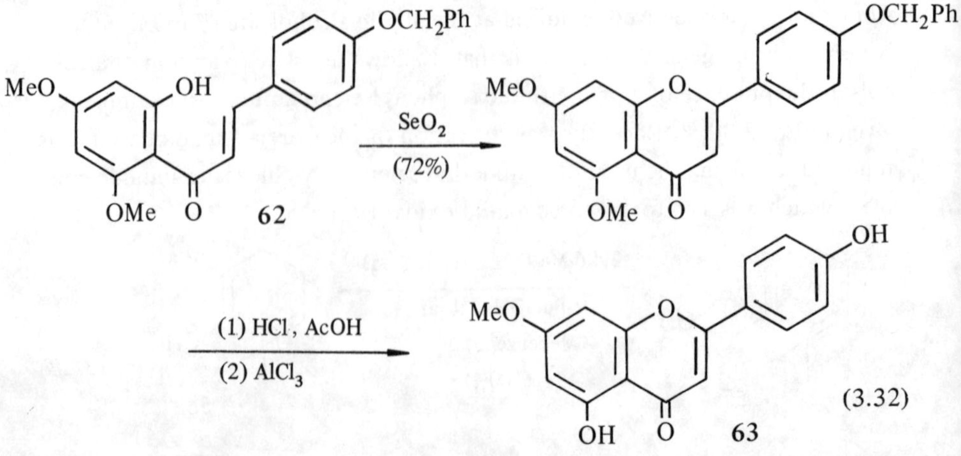

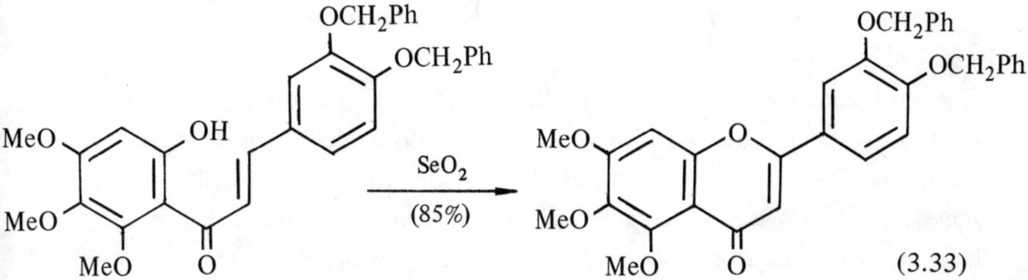

SeO_2 (85%)

(3.33)

SeO_2, t-BuOH, Δ (80%)

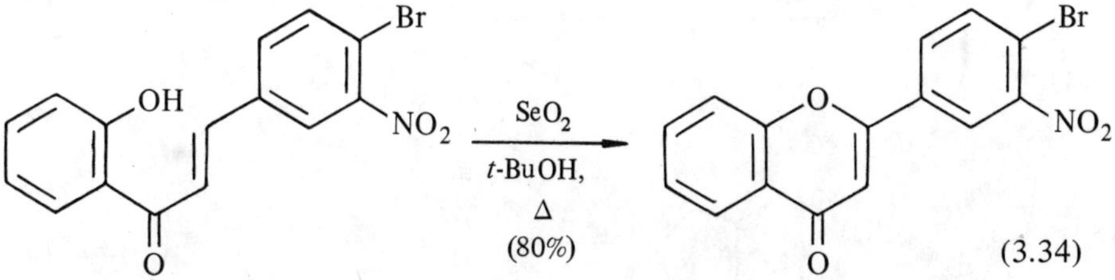

(3.34)

Use of this method made possible the transformation of **64** to **65**, which was oxidized to O-methylkidamycinone (**66**), as shown in Eq. 3.34a.[25]

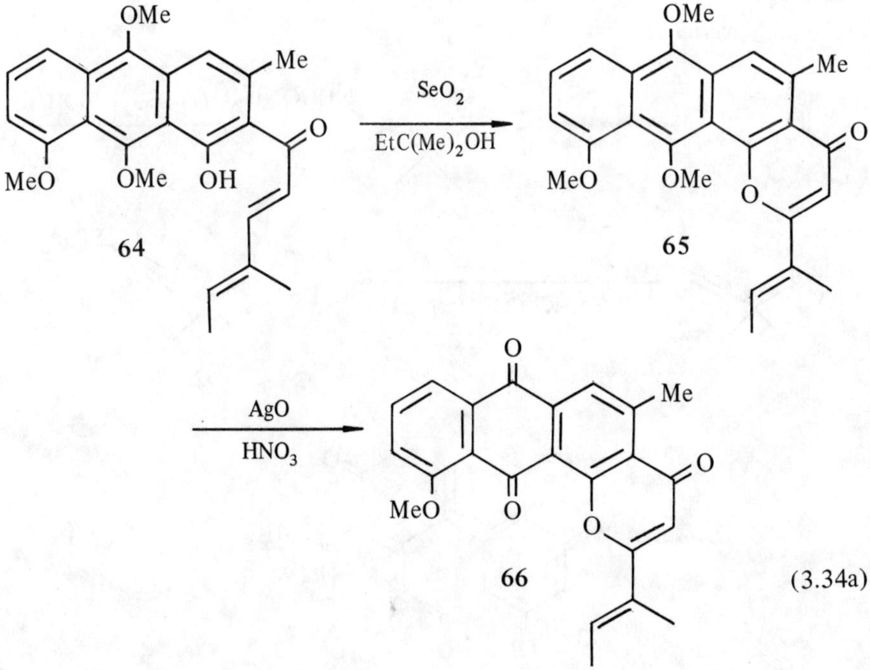

SeO_2, $EtC(Me)_2OH$

64 **65**

AgO, HNO_3

66

(3.34a)

3.4 Transformation of ketones to alkynes via selenadiazoles

When ketone semicarbazones (**68**) are treated with selenium dioxide in acetic acid, the product is a 1,2,3-selenadiazole (**69**), which upon pyrolysis gives the alkyne (**70**, Eq. 3.35).[26] Overall, the series of reactions constitutes deoxygenation–dehydrogenation of a ketone (**67**) to the corresponding alkyne (**70**); it has been used to make several alkynes, including cyclic alkynes.[27]

$H_2NNHCONH_2$; $N-NHCONH_2$; SeO_2; Se; Δ

67 **68** **69** **70**

(3.35)

Recently, in an elegant synthesis of the antibiotic resistomycin (**73**), outlined in Eq. 3.36,[28] acetylenic derivative 72 was prepared from ethyl acetoacetate (71) in 22% overall yield.

EtO; $N-NHCONH_2$; (1) SeO_2, AcOH; (2) Δ; 71

(1) Base; (2) Me_3SiCl (22%, from 71); $SiMe_3$; 72

HO; Me; OH; OH; O; OH; 73

(3.36)

3.5 Dehydrogenations of N-*heterocycles*

Elemental selenium and sometimes selenium dioxide have been used in the past to fully dehydrogenate polycyclic systems, forming fused aromatic hydrocarbons.[29] Aromatization seems to occur more readily with heterocyclic systems. In a recent study, treatment of **74** with selenium dioxide in refluxing dioxane gave a mixture of **75**, **76**, **77** and **78** (Eq. 3.37).[30] The major product (**77**), obtained in 32% yield, is a naturally occurring indole alkaloid, as is **78**, obtained in 13% yield.

COOMe, NH, N, H, SeO_2, R, N, N, H

74

75: R = COOMe
76: R = H

SeO_2, R, N, N, H, O

77: R = COOMe
78: R = H

(3.37)

The same workers also prepared the alkaloid canthin-6-one (**81**) from *N*-benzyltryptamine (**79**) in two steps, employing the same dehydrogenation procedure, as shown in Eq. 3.38.[30] Similarly, derivative **82**, which posseses high affinity for the benzodiazepin receptor, was prepared from **80** in excellent yield.[30]

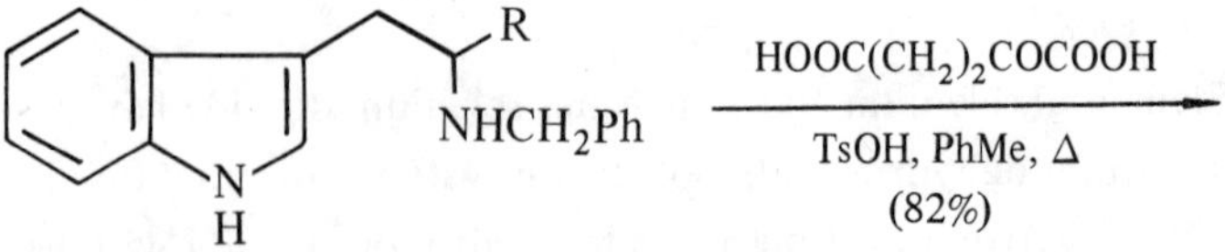

$$\xrightarrow[\text{TsOH, PhMe, }\Delta\ (82\%)]{HOOC(CH_2)_2COCOOH}$$

79: R = H

80: R = COOMe

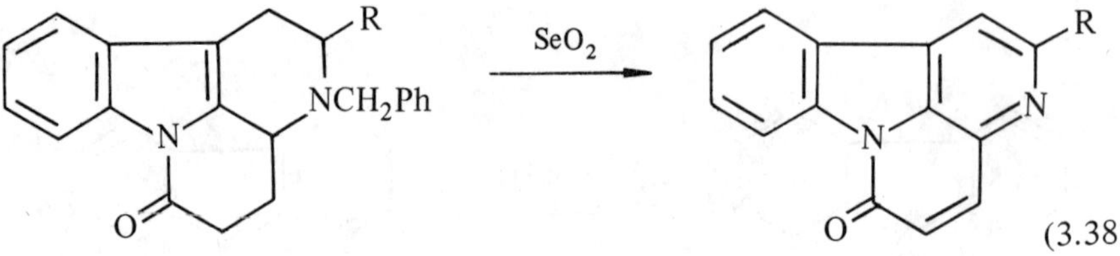

$$\xrightarrow{SeO_2}$$

(3.38)

81: R = H (35%)

82: R = COOMe (66%)

A milder reagent for the dehydrogenation of substituted tetrahydropyridine systems is the Se(IV) derivative diphenylselenium bis(trifluoroacetate), $Ph_2Se(OCOCF_3)_2$, introduced recently by Marino.[31] This compound, called a selenurane, is prepared from diphenyl selenoxide (Ph_2SeO) and trifluoroacetic anhydride, $(CF_3CO)_2O$, (Eq. 3.39).[31] Reactions with this reagent proceed even at room temperature and it can be used to prepare partially dehydrogenated or fully aromatized products. For example, **83** gives **84** in 80% yield (Eq. 3.40).[31] Treatment of **85** with excess reagent gives paraveraldine (**86**), formed by further oxidation of the methylene group to a carbonyl (Eq. 3.41).[31] Also, **87** was dehydrogenated efficiently to **88** with 3 eq of the reagent at room temperature (Eq. 3.42).[31] Tertiary amines are oxidized to the corresponding iminium species, which can be quenched with potassium cyanide to give the cyano-substituted derivatives such as **89**, in high yield (Eq. 3.43).[31]

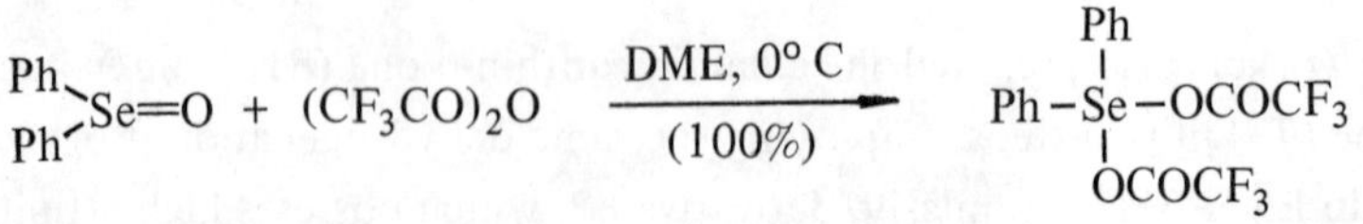

$$Ph_2Se{=}O + (CF_3CO)_2O \xrightarrow[(100\%)]{\text{DME, }0°\text{ C}} Ph_2Se(OCOCF_3)_2 \qquad (3.39)$$

$$\text{83} \xrightarrow[\text{(80\%)}]{Ph_2Se(OCOCF_3)_2\ (2\ eq)} \text{84} \quad (3.40)$$

$$\text{85} \xrightarrow[\text{(75\%)}]{Ph_2Se(OCOCF_3)_2\ (6\ eq)} \text{86} \quad (3.41)$$

$$\text{87} \xrightarrow[\text{(70\%)}]{Ph_2Se(OCOCF_3)_2\ (3\ eq)} \text{88} \quad (3.42)$$

$$\text{MeO, MeO, NMe} \xrightarrow[\text{(2) KCN (90\%)}]{\text{(1) } Ph_2Se(OCOCF_3)_2} \text{89 (CN)} \quad (3.43)$$

Diphenyl selenoxide, Ph_2SeO, is a mild reagent for the oxidation of catechols to *o*-quinones, as well as for several other oxidations.[32] Treatment of L-adrenaline (**90**) with diphenyl selenoxide gave adrenochrome (**91**), as indicated in Eq. 3.44.[32]

$$\text{90} \xrightarrow[\text{HCOOH–MeOH (72\%)}]{PhSe(O)Ph} \text{91} \quad (3.44)$$

REFERENCES

1. For reviews, see ref. 1, Chapter 2. See also § 2.4.
2. (a) Corey, E. J., and Schaefer, J. P., *J. Am. Chem. Soc.,* **82**, 918 (1960); (b) Schaefer, J. P., *J. Am. Chem. Soc.,* **84**, 713,717 (1962); (c) Jerussi, R. A., and Speyer, D., *J. Org. Chem.,* **31**, 3199 (1966).
3. Sharpless, K. B., and Gordon, K. M., *J. Am.,Chem. Soc.,* **98**, 300 (1976).
4. (a) Meystre, C., Frey, H., Voser, W., and Wettstein, A., *Helv. Chim. Acta,* **39**, 734 (1956); (b) Bernstein, S., and Littell, R., *J. Am. Chem. Soc.,* **82**, 1235 (1960).
5. Branca, S. J., Lock, R. L., and Smith, A. B., III, *J. Org. Chem.,* **42**, 3165 (1977).
6. (a) Branca, S. J., and Smith, A. B., III, *J. Am. Chem. Soc.,* **100**, 7767 (1968); (b) Smith, A. B., III, and Pilla, N. N., *Tetrahedron Lett.,* **21**, 4961 (1980).
7. Viswanatha, V., and Rao, G. S. K., *Tetrahedron Lett.,* 247 (1974).
8. (a) Nagaoka, H., Schmid, G., Iio, H., and Kishi, Y., *Tetrahedron Lett.,* **22**, 899 (1981); (b) Iio, H., Nagaoka, H., and Kishi, Y., *J. Am. Chem. Soc.,* **102**, 7965 (1980).
9. Tang, C., and Rapoport, H., *J. Am. Chem. Soc.,* **94**, 8615 (1977).
10. Lee, Y. W., Lee, E., and Nakanishi, K., *Tetrahedron Lett.,* **21**, 4323 (1980).
11. Koft, E. R., and Smith, A. B., III, *J. Am. Chem. Soc.,* **104**, 2659 (1982).
12. Barnes, C. S., and Barton, D. H. R., *J. Chem. Soc.,* 1419 (1953).
13. (a) Wiesner, K., Jirkovsky, J., Fishman, M., and Williams, C. A. J., *Tetrahedron Lett.,* 1523 (1967); (b) Wiesner, K., and Jirkovsky, J., *Tetrahedron Lett.,* 2077 (1967); (c) Wiesner, K., and Poon, L., *Tetrahedron Lett.,* 4937 (1967).
14. Marx, J. N., Cox, J. H., and Norman, L. R., *J. Org. Chem.,* **37**, 4489 (1972).
15. (a) Barton, D. H. R., Lester, D. J., and Ley, S. V., *J. C. S. Perkin I,* 2209 (1980); (b) Barton, D. H. R., Lester, D. J., and Ley, S. V., *J. C. S. Chem. Comm.,* 130 (1978).
16. (a) Barton, D. H. R., Godfrey, C. R. A., Morzycki, J. W., Motherwell, W. B., And Ley, S. V., *J. C. S. Perkin I,* 1947 (1982); (b) Barton, D. H. R., Morzycki, J. W., Motherwell, W. B., and Ley, S. V., *J. C. S. Chem. Comm.,* 1044 (1981).
17. (a) Barton, D. H. R., Hui, R. A. H. F., Ley, S. V., and Williams, D. J., *J. C. S.*

Perkin I, 1919 (1982); (b) Barton, D. H. R., Hui, R. A. H. F., Lester, D. J., and Ley, S. V., *Tetrahedron Lett.,* 3331 (1979).

18. (a) Back, T. G., *J. Org. Chem.,* **46**, 1442 (1981); (b) Back, T. G., *J. C. S. Chem. Comm.,* 278 (1978).
19. Barton, D. H. R., Brewster, A. G., Hui, R. A. H. F., Lester, D. J., Ley, S. V., and Back, T. G., *J. C. S. Chem. Comm.,* 952 (1978).
20. (a) Kuwajima, I., Shimizu, M., and Urabe, H., *J. Org. Chem.,* **47**, 837 (1982); (b) Shimizu, M., and Kuwajima, J., *Tetrahedron Lett.,* 2801 (1979).
21. Dhar, D. N., *The Chemistry of Chalcones and Related Compounds,* Wiley-Interscience, New York (1981).
22. Mahal, H. S., and Venkataraman, K., *J. Chem. Soc.,* 569 (1936).
23. Matsumura, S., Kunii, T., and Matsumura, A., *Chem. Pharm. Bull.,* **21**, 2757 (1973).
24. Huang, D. L., Weng, T., and Chen, F. C., *J. Heterocyclic Chem.,* 7, 1189 (1970).
25. Hauser, F. M., and Rhee, R. P., *J. Am. Chem. Soc.,* **101**, 1628 (1979).
26. Lalezari, I., Shatiee, A., and Yalpani, M., *J. Org. Chem.,* **36**, 2836 (1971).
27. (a) Lalezari, I., Shafiee, A., and Yalpani, M., *J. Heterocyclic Chem.,* **9**, 1411, (1972); (b) Meier, H., and Menzel, I., *J. C. S. Chem. Comm.,* 1059 (1971); (c) Golgolab, H., and Lalezari, I., *J. Heterocyclic Chem.,* **12**, 801 (1975); (d) Meier, H., Echter, T., and Peterson, H., *Ang. Chem. Int. Ed. Engl.,* **17**, 942 (1979).
28. Keay, B. A., and Rodrigo, R., *J. Am. Chem. Soc.,* **104**, 4725 (1982).
29. Fu, P. P., and Harvey, R. G., *Chem. Rev.,* 78, 317 (1978).
30. (a) Cain, M., Campos, O., Guzman, F., and Cook, J. M., *J. Am. Chem. Soc.,* **105**, 907 (1983); (b) Campos, O., DiPierro, M., Cain, M., Mantei, R., Gawishi, A., and Cook, J. M., *Heterocycles,* **14**, 975 (1980).
31. Marino, J. P., and Larser, R. D., Jr., *J. Am. Chem. Soc.,* **103**, 4642 (1981).
32. Balenovic, K., Bregant, N., and Perina, I., *Synthesis*, 172 (1973).
33. Takaki, K., Yasumura, M., and Negoto, K., *J. Org. Chem.,* **48**, 54 (1983).

CHAPTER 4

Organoselenium-mediated Olefinations

A very important property of aryl alkyl selenoxides having a β-hydrogen, such as 2, is that they undergo *syn* elimination to give olefins (3) under relatively mild conditions. Furthermore, the selenoxides (2) can be easily prepared by oxidation of the corresponding selenides (1) and so the overall conversion of 1 to 3 (Eq. 4.1) constitutes a very useful method for the introduction of olefinic bonds.

SeAr [O] SeAr O H −ArSeOH (4.1)

H

1 2 3

The first example of selenoxide *syn* elimination was observed by Jones[1] during an attempt to prepare a steroidal selenoxide optically pure at selenium. Although the resolution was successful, the selenoxides obtained decomposed rapidly at room temperature, forming the olefin in high yield, as shown in Eq. 4.2. Shortly after Jones' report, another selenoxide elimination was observed during oxidation studies on a selenocysteine derivative, outlined in Eq. 4.3.[2]

H O_3 H 25° C H

H −78° C H (>95%) H

H H H

Ph Se Ph Se=O (4.2)

$$Ph_2CHSe\text{-}CH_2CH(NHCOOCHPh_2)COOCHPh_2 \xrightarrow[H_2O_2]{NaIO_4 \text{ or}} CH_2{=}C(NHCOOCHPh_2)COOCHPh_2 \quad (4.3)$$

The synthetic potential of the reaction was first realized by Sharpless[3–7] and by Reich,[8–12] who developed it into a useful synthetic method. Sharpless' and Reich's monumental contributions initiated a stream of fruitful research in organoselenium chemistry, with many applications.[7, 12–17]

The only requirement for the synthesis of olefins (3) is the preparation of the selenides (1) from available substrates. This has been achieved by taking advantage of an existing functionality such as carbonyl, olefin, epoxide, etc., or by transformation of other selenides. A major advantage of the arylseleno group is that it can act either as an electrophilic species ($ArSe^{\oplus}$) or as a nucleophilic one ($ArSe^{\ominus}$). Thus, the nucleophilic substrates, such as carbanions or olefins, are treated with electrophilic organoselenium reagents (ArSeX), while the electrophilic substrates such as epoxides or other good leaving groups are treated with nucleophilic organoselenium reagents (ArSeM). All of these methods for the introduction of an olefinic bond into a molecule will be examined in this chapter.

The most commonly used arylseleno group is the unsubstituted one (PhSe). Several reagents are now commercially available or are readily prepared for the introduction of this group.[3–64] Table 4.1 shows the various electrophilic and nucleophilic reagents carrying the PhSe group. Some other organoselenium groups used are the *o*-nitrophenylseleno group (*o*-$O_2NC_6H_4Se$),[29, 30, 52] the methylseleno group (MeSe),[23, 65, 66] and the 2-pyridylseleno group (2-Py-Se).[67]

The oxidation of selenides (1) to the selenoxides (2) can be done by one of several methods, depending on the substrate. The most commonly used methods of oxidation are shown in Table 4.2.

Usually, the selenides are oxidized much faster than other functional groups and most of these oxidations can be done at low temperature. Every reagent, however, has its own advantages and limitations, such as availability, compatibility, reactivity, solubility, etc.[11, 12, 14, 33, 34]

The selenoxide *syn* elimination usually takes place at room temperature or in refluxing carbon tetrachloride or hexane. The addition of an amine (diisopropylamine, triethylamine, pyridine, etc.) prior to the elimination, is recommended, in order to avoid undesirable side reactions , such as epoxidation and addition of the by-product selenenic acid to olefins.[34]

Table 4.1. Organoselenium reagents for the introduction of the phenylseleno group (PhSe).

Reagent	*Active species*	*Ref.*
A. Electrophilic	*PhSe⊕X⊖*	
PhSeSePh	PhSeSePh	5
PhSeCl	PhSeCl	5–9, 11, 18–20
PhSeSePh–Br_2	PhSeBr	5–9, 11, 21–23
PhSeSePh–I_2	PhSeI	24
PhSeBr–CF_3COOAg	$PhSeOCOCF_3$	10, 25, 26
PhSeBr–KOAc–HOAc	PhSeOAc	6
PhSeCl(Br)–ROH	PhSeOR	6, 27
PhSeCN–$CuCl_2$–ROH	PhSeOR	28
PhSeCN–*n*-Bu_3P	$PhSe^nBu_3P^{\oplus}CN^{\ominus}$	29–31
PhSeCN–$SnCl_4$	PhSeCN	32
PhSeCN–$CuCl_2$–H_2O	PhSeOH	28
PhSeSePh–H_2O_2	PhSeOH	33, 34
PhSeOOH–H_3PO_2	PhSeOH	35
PhSeCl–H_2O	PhSeOH	36
PhSeSePh–*t*-BuOOH or $(PhSeO)_2O$	PhSeOSePh	37
PhSeSePh–Br_2–$(Bu_3SnO)_2$	$PhSeOSnBu_3$	228
PhSe—N(succinimide) (*N*-PSS)	*N*-PSS	38, 39
PhSeN(phthalimide) (*N*-PSP)	*N*-PSP	39, 40
N-PSS or *N*-PSP–H_2O	PhSeOH	41
PhSeCl(Br)–Me_2NH	$PhSeNMe_2$	42, 43
PhSeCl–Et_2NH	$PhSeNEt_2$	42, 224

Table 4.1—*continued*

Reagent	*Active species*	*Ref.*
PhSeCl–MeCN, CF_3SO_3H, H_2O	PhSeNHCOMe	44
PhSeBr–$HgCl_2$, $AgNO_2$	$PhSeNO_2$	45
$PhSeSO_2Ar$	$PhSeSO_2Ar$	46–49
B. Nucleophilic	*$PhSe^{\ominus}M^{\oplus}$*	
PhSeSePh–$NaBH_4$	PhSeNa	3,4,50,51
PhSeCN–$NaBH_4$	PhSeNa	52
PhSeSePh–Na	PhSeNa	53, 54
PhSeSePh–NaOH, R_4NCl	PhSeNa	55
PhSeH–NaH	PhSeNa	29
PhSeH–*n*-BuLi	PhSeLi	56
PhSeH–EtOTl	PhSeTl	57
PhSeSePh–H_3PO_2	PhSeH	56
$(PhSe)_3B$	PhSeH	58
$PhSeSiMe_3$–KF, 18-C-6	PhSeK	59
$PhSeSiMe_3KI$, 18-C-6	PhSeK	60
$PhSeSiMe_3$–catalyst	$PhSeSiMe_3$	59–64

Table 4.2. Reagents for oxidizing selenides to selenoxides.

Reagent	*Ref.*
H_2O_2	3–5, 8, 11
O_3	1, 11, 34, 68
$NaIO_4$	5, 8, 11, 25
AcOOH	5
mCPBA	8, 34, 69
NBS or NCS	70, 71
t-BuOCl	71, 72
t-BuOOH	33
t-BuOOH–Al_2O_3	222
TsNClNa	7
$PhSO_2N(O)CHAr$	73

4.1 *Transformation of carbonyl compounds to their α,β-unsaturated derivatives*

Formation of the α,β-unsaturated carbonyl group is a very important synthetic process. This type of functionality is not only quite useful in synthetic intermediates, due to its versatility, but it also exists as such in numerous natural products. Synthetic precursors are usually the corresponding saturated carbonyl compounds, and traditional methodology includes α-bromination–dehydrobromination and direct dehydrogenation with various reagents (Chap. 3). Application of organoselenium chemistry to this area was first explored by Sharpless[5] and Reich.[8, 11] As a result, a powerful synthetic method has evolved, which makes this type of transformation the most widely used application of organoselenium reagents.

The method consists in treating the enolic form of the carbonyl substrate (**4**) with an electrophilic carrier of the phenylseleno group (*e.g.*, PhSeSePh, PhSeCl, PhSeBr, etc.) and subsequently oxidizing the resulting α-phenylseleno carbonyl compound (**5**) to the corresponding selenoxide, which undergoes syn elimination to form the α,β-unsaturated carbonyl product (**6**, Eq. 4.4). The mildness, efficiency and regioselectivity of the method make it suitable for use with delicate substrates of the type often encountered in natural products synthesis. As shown below, the method has been used to prepare α,β-unsaturated ketones, aldehydes, esters, lactones, nitriles and lactams.

OR, H (4) —PhSeX→ O, SePh, H (5) —(1) [O], (2) Δ→ O (6) (4.4)

R = H, Li, Na, K, Cu, Zr, Al, Ac, $SiMe_3$ or Me

4.1.1 *Synthesis of α,β-unsaturated ketones*

Highly enolizable ketones can be selenenylated directly by treating them with phenylselenenyl chloride (PhSeCl), but not with phenylselenenyl bromide (PhSeBr) or diphenyl diselenide (PhSeSePh), in ethyl acetate solution.[5] By using this method, cholestanone (**7**) was transformed to enone **8** in high yield and regiospecifically,as shown in Eq. 4.5.[5] Furthermore, the method proved to be very useful for the introduction of the cyclopentenone moiety in several

(1) PhSeCl, EtOAc
(2) H_2O_2
(84%)

7 8 (4.5)

pseudoguaianolides, as in the conversion of damsin (**9**) to ambrosin (**10**), shown in Eq. 4.6,[74] and in the preparation of aromatin (**12**) from dihydroaromatin (**11**, Eq. 4.7).[75]

(1) PhSeCl, EtOAc
(2) $NaIO_4$, *t*-BuOH
(41%)

9 10 (4.6)

(1) PhSeCl–EtOAc–HCl
(2) $NaIO_4$, THF–H_2O
(75%)

11 12 (4.7)

If the enol is formed regiospecifically in one direction due to steric factors, then the selenenylation takes place exclusively towards that direction. This was the case in Danishefsky's synthesis of quadrone (**16**), outlined in Eq. 4.8, in which intermediate **13** was transformed to enone **15**, presumably via selenylation of enol **14**.[76] Noteworthy is the use of the olefinic bond in this synthesis as a means of achieving the introduction of the hydroxymethyl group in the less favored position.

O HO
HOOC HOOC
13 14

(1) PhSeCl, EtOAc
(2) H_2O_2–pyr (87%)

O (1) LDA O
HOOC (2) CH_2O (62%) HOOC
HO 15 (4.8)

(1) H_2–Pd/C
(2) Δ ($-H_2O$)

O
O O
16

A more general method for the selenenylation of ketones is the one introduced by Reich,[8,11] which consists in treating the substrate with a strong base (*e.g.*, lithium diisopropyl amide, LDA) at low temperature and quenching the resulting lithium enolate with phenylselenenyl chloride or bromide (but not with diphenyl diselenide).This method was used in an efficient conversion of exaltone (**17**) to muscone (**18**, Eq. 4.9),[77] as well as in Stork's synthesis of prostaglandin A_2 (**20**, Eq. 4.10),[78] and in Paquette's synthesis of isocomene (**21**, Eq. 4.11).[79]

O O O
(1) LDA–PhSeBr LiCuMe$_2$
(2) H_2O_2 (79%, overall)
17 18
(4.9)

(4.10)

(4.11)

As shown in the conversion of **19** to **20**, the method has a high degree of regioselectivity, which is achieved by the formation of the kinetic enolate of the ketone with the strong and sterically hindered base diisopropyl lithium amide (LDA). This particular advantage of the method has been utilized on several occasions, as in the synthesis of α-cardinol (**22**, Eq. 4.12),[80] coriolin (**23**, Eq. 4.13),[81, 82] and pentalene (**24**, Eq. 4.14).[83]

(1) LDA–PhSeBr
(2) H_2O_2 (75%)

$h\nu$, HOAc (40%)

OAc H OH H H

22

(4.12)

(1) LDA–PhSeCl
(2) H_2O_2
(3) HOAc (50%)

H H OTHP H H OH HO H H OH

23

(4.13)

(1) LDA–PhSeCl
(2) H_2O_2 (64%)

$LiCuMe_2$
(87%)

(1) LDA–PhSeCl
(2) H_2O_2 (57%)

$(Ph_3P)_3RhCl$
Et_3SiH

H_2NNH_2–K_2CO_3
Δ

24

(4.14)

A stereochemical feature of the method is that the phenylseleno group is introduced on the less hindered side of a sterically hindered substrate. This is important in cases where there is only one β-hydrogen, since the phenylseleno group must be placed *cis* to this hydrogen in order for *syn* elimination to occur. For example, in Semmelhack's synthesis of illudol (**28**), by choosing the correct sequence of reagents, ketone **25** could be converted either to selenide **26** or to **29** (Eq. 4.15).[84] Subsequently, **26** underwent oxidation–*syn* elimination to give the unsaturated product (**27**), which was converted to illudol (**28**). The other isomer (**29**) gave α-formyl ketone **30** on oxidation with hydrogen peroxide.

(4.15)

Spirodienones such as 32 can also be prepared by this method from the corresponding spiroenones (31, Eq. 4.16), without any rearrangement to non-spiro derivatives.[85]

(1) LDA–PhSeCl

(2) H_2O_2 (83%)

31 32

(4.16)

Enolates resulting from the Michael addition of cuprates to enones react with electrophilic PhSeX reagents, producing an overall β-alkylation–α-phenylselenenylation in one step.[9,11] This type of transformation was used in a synthesis of modhephene (33) by Oppolzer, highlighted in Eq. 4.17.[86]

(1) $CH_2{=}CHCH_2CH_2MgBr$ $CuBr \cdot Me_2S$

(2) PhSeBr (64%)

SePh

H_2O_2–pyr (84%)

Δ (76%)

H

H_2–Pd/C (98%)

H

(1) $CH_2{=}PPh_3$

(2) TsOH (66%)

H

33

(4.17)

If the β-position of the enone is monosubstituted, the sequence (1) cuprate addition, (2) selenenylation and (3) selenoxide elimination, leads to overall β-alkylation of the enone. This process was used in a synthesis of the biologically active diterpene nagilactone F (**34**), as shown in Eq. 4.18.[87]

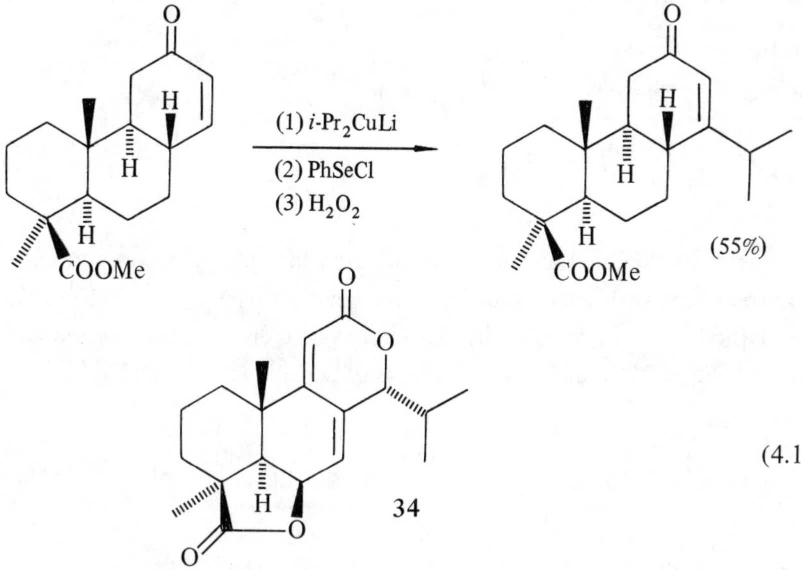

(1) i-Pr_2CuLi
(2) PhSeCl
(3) H_2O_2

(55%)

34

(4.18)

Similar reaction sequences involving conjugate addition of vinyl zirconium or acetylenic aluminum reagents to enones, were reported by Schwartz.[229]

α-Phenylselenation of acyclic ketones, followed by oxidative *syn* elimination, leads to *trans*-enones. An example is shown in Eq. 4.19, taken from Cava's approach to the aklavinones.[88] The selenation step in this case was done under acidic conditions.

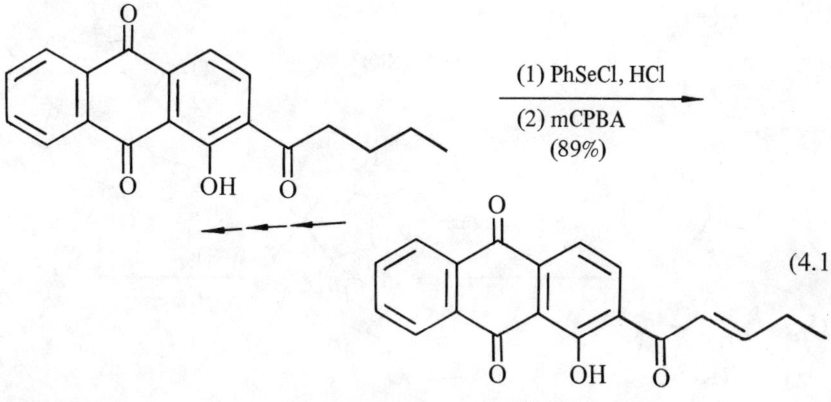

(1) PhSeCl, HCl
(2) mCPBA
(89%)

(4.19)

An alternative method for the preparation of enones via α-phenylseleno ketones involves treatment of preformed enol acetates with phenylselenenyl trifluoroacetate (formed from phenylselenenyl bromide and silver trifluoroacetate).[10,11,25] This is illustrated in the preparation of cyclohexanone (**35**), as shown in Eq. 4.20.[25]

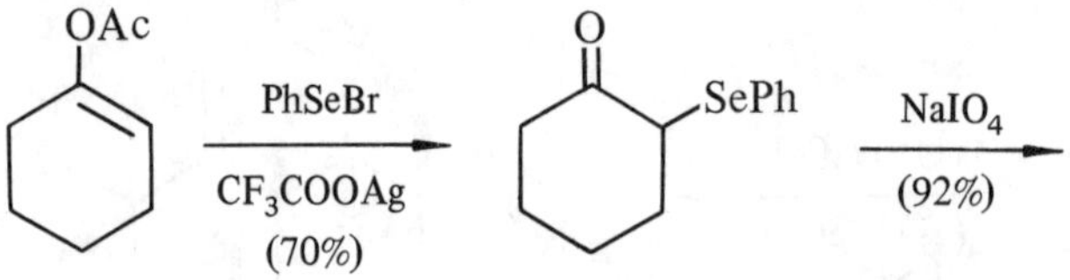

(4.20)

Trimethylsilyl enol ethers are also converted to α-phenylseleno ketones upon treatment with phenylselenenyl chloride[89] or phenylselenenyl bromide.[90] The method was used by Danishefsky to convert diene **36** to phenylselenodiene **37** (Eq. 4.21)[91] and in his synthesis of coriolin (**23**, Eq. 4.22).[92]

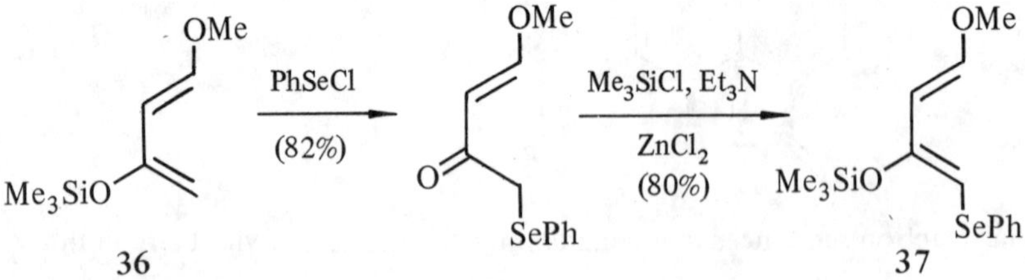

(4.21)

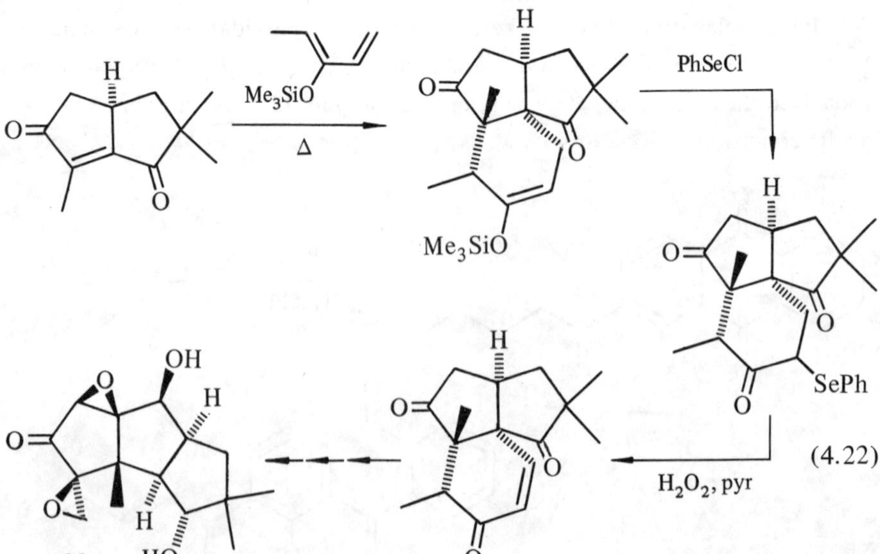

(4.22)

A similar reaction is observed with methyl enol ethers on treatment with phenylselenenyl bromide–silver trifluoroacetate, as in Semmelhack's synthesis of deoxyfrenolicin (**41**), shown in Eq. 4.23.[93]

MeO O MeOOC O **38** — (1) $(MeO)_3CH$, TsOH; (2) Δ (−MeOH) → MeO O MeO MeOOC — $PhSeBr{-}AgOOCCF_3$ → MeO O O SePh MeOOC **39** — O_3 (62%) → MeO O OH MeOOC **40** — (1) Jones oxidation; (2) BBr_3; (3) KOH → HO O O O COOH **41**

(4.23)

As shown in the conversion **38**→**39**→**40**, cyclohexenones without a disubstituted ring carbon (hydroaromatic rings) are converted to the corresponding phenols upon selenenylation–selenoxide elimination. An alternative reaction sequence for this transformation involves the attachment of the PhSe group with PhSeCl–LDA and its oxidative removal with *m*-chloroperbenzoic acid in the presence of 3,5-dimethoxyaniline.[94] The method is illustrated in the conversion of **42** to **43** (Eq. 4.24),[94] and in the synthesis of the tetrahydrocannabinoid derivative **44**, as shown in Eq. 4.25.[94]

$$\xrightarrow[\text{(45\%)}]{\text{LDA}-\text{PhSeCl}} \quad \xrightarrow[\text{(73\%)}]{\text{mCPBA},\ 3,5\text{-(MeO)}_2\text{C}_6\text{H}_3\text{NH}_2} \quad (4.24)$$

42 → 43

$$\xrightarrow[\text{(76\%)}]{\text{LDA}-\text{PhSeCl}} \quad \xrightarrow[\text{(66\%)}]{\text{mCPBA},\ 3,5\text{-(MeO)}_2\text{C}_6\text{H}_3\text{NH}_2} \quad (4.25)$$

44

Ketones, as well as aldehydes, can also be selenenylated directly with selenium dioxide and diphenyl diselenide, in the presence of an acid catalyst.[223]

A different arylseleno group, namely the 2-pyridylseleno group, was introduced recently for the preparation of enones and enals.[67] 2-Pyridylseleno is considered to be a better leaving group than phenylseleno. An example of the use of 2-pyridylseleno is shown in Eq. 4.26.[67]

$$\xrightarrow[\text{HCl (84\%)}]{\text{2-PySeSePy-2, Br}_2} \quad \xrightarrow[\text{(100\%)}]{\text{O}_3,\ \text{Et}_2\text{NH}} \quad (4.26)$$

4.1.2 Synthesis of α,β-unsaturated aldehydes

Direct selenenylation of aldehydes with phenylselenenyl chloride is possible.[5] In some cases, however, the product of the reaction is the α,β-unsaturated aldehyde, formed by nonoxidative elimination of the elements of PhSeH, instead of the α-phenylseleno aldehyde.[95,96] This was observed in Goldsmith's synthesis of waburganol (**45**), outlined in Eq. 4.27,[95] and in Schlessinger's synthesis of eriolanin (**46**, Eq. 4.28).[96]

NaH, HCOOEt (91%) — CHOH — PhSeCl, pyr (100%)

CHO, SePh + CHO — H_2O_2

(1) $HO(CH_2)_3OH$, $H^{\oplus}$
(2) MeLi

(4.27)

HO — OH — O — O — (1) $MeOOCNSO_2NEt$, Et_3N (2) OsO_4 — HO — O — O

(1) [O]
(2) TsOH

CHO — OH — CHO

H

45

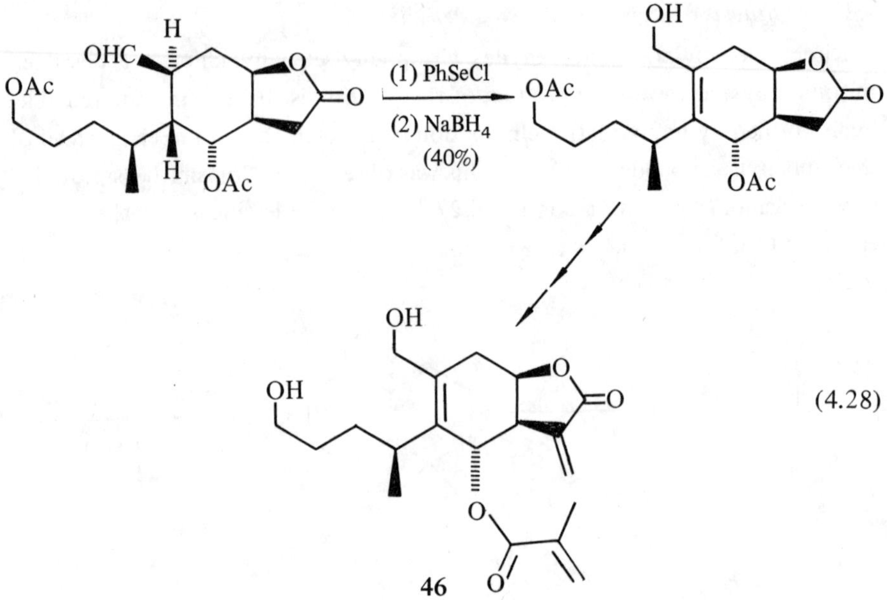

(1) PhSeCl
(2) $NaBH_4$
(40%)

46

(4.28)

An alternative procedure for the direct selenenylation of aldehydes involves treatment with *N,N*-diethylbenzeneselenamide ($PhSeNEt_2$).[224]

α-Phenylseleno aldehydes (**49**) can be prepared by a method developed by Nicolaou which consists in treating the corresponding methyl enol ethers (**48**) with phenylselenenyl chloride (Eq. 4.29).[97] The methyl enol ethers are readily available from the lower homolog carbonyl compounds (**47**) by a Wittig reaction. The overall conversion of **47** to the α,β-unsaturated aldehydes (**50**) is a net homologation–dehydrogenation process. It was used in Nicolaou's synthesis of the biologically active carbocyclic analog of thromboxane A_2 (**51**), shown in Eq. 4.30.[98]

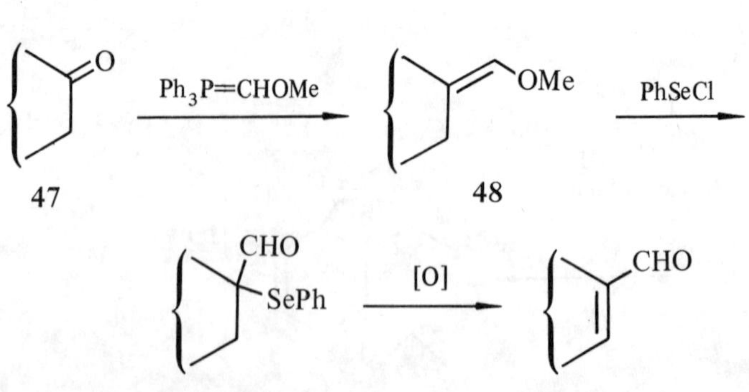

$Ph_3P{=}CHOMe$ → PhSeCl →

47 48

[O] →

49 50

(4.29)

(1) $Ph_3P{=}CHOMe$
(2) PhSeCl
(62%)

CHO
SePh

mCPBA
(88%)

CHO

(1) LiCu
OSi^{t}BuMe$_2$
(2) K_2CO_3
(56%)

CHO
OSi^{t}BuMe$_2$

(1) $Ph_3P{=}CHOMe$
(2) $Hg(OAc)_2$–KI

CHO
OSi^{t}BuMe$_2$

(1) $Ph_3P{=}CH(CH_2)_3COO^{\ominus}$
(2) CH_2N_2
(3) $H_3O^{\oplus}$
(4) LiOH

COOH
OH
51

(4.30)

A similar method was used in a recent synthesis of flourensic acid (**52**), shown in Eq. 4.31.[99]

$Ph_2P(O)CH_2OMe$–LDA (78%)

(1) $PhSeCl$–$AgOOCCF_3$
(2) HCl
(3) H_2O_2
(50%)

Jones oxidation (78%)

52

(4.31)

In a recent synthesis of domoic acid (**56**) using this method (Eq. 4.32) it was found that oxidation–*syn* elimination of α-phenylselenoaldehyde **53** gives varying ratios of the *E*-enal (**54**) and the desired *Z*-enal (**55**), depending on the oxidant used.[100] Ozonolysis followed by treatment with triethylamine gave a 10:1 mixture of **54** and **55** in 30% yield, while bromination with NBS followed by treatment with sodium acetate gave a 67% yield of a 1:2 mixture of **54** and **55**. Other oxidants, *e.g.* hydrogen peroxide and sodium periodate, did not work at all.

$Ph_3P{=}CHOMe$

PhSeCl (90%)

(4.32)

Boc—N CHO SePh MeOOC COOMe **53** [O] → CHO COOMe **54** + CHO COOMe **55** → HN HOOC HOOC H COOH **56**

(4.32)

Preparation of α,β-unsaturated aldehydes (**50**) from the corresponding α,β-saturated precursors (**57**) can be accomplished by converting the aldehydes **57** to their enamines (**58**) which upon treatment with phenylselenenyl chloride give the α-phenylselenoaldehydes (**49**), which are transformed in turn to the products (**50**) by oxidative elimination using mCPBA or sodium periodate (Eq. 4.33).[101] This method was developed by Williams; as shown in the conversion of **59** to **60** (Eq. 4.34), the selenoxide elimination occurs with some regioselectivity.[101]

CHO **57** Piperidine → N **58** PhSeCl →

CHO SePh **49** [O] → CHO **50**

(4.33)

Piperidine (97%)

(1) PhSeCl
(2) $NaIO_4$
(50%)

59 CHO → 60 CHO (4.34)

Williams also showed that α-phenylselenoaldehydes (**49**) can be reduced to the corresponding alcohols (**61**), which lead to allylic alcohols (**62**, Eq. 4.35) without the intermediacy of the aldehydes (**50**) which might result in the epimerization of neighboring chiral centers.

CHO, SePh (**49**) —[H]→ OH, SePh (**61**) —mCPBA→ OH (**62**) (4.35)

This was advantageous in the conversion of **63** to **64** (Eq. 4.36), an intermediate in William's synthesis of the macrolide antibiotic milbemycin β_3.[102]

Br, CHO (**63**) → Br, OH (**64**)

(1) Piperidine
(2) PhSeCl
(3) $LiAl(O^tBu)_3H$
(4) mCPBA
(76%)

(4.36)

Selenenylation of aldehyde enolates is not an efficient process, but trapping the enolate as a trimethylsilyl enol ether and subsequent treatment with phenylselenenyl chloride affords the desired α-phenylselenoaldehydes.[90] This approach was used by Paquette for the preparation of α-phenylselenoaldehydes **66** and **67**, which were formed in a 9:1 ratio from **65** (Eq. 4.37).[103]

OH (1) KH (2) Me_3SiCl → 65 ($OSiMe_3$) → PhSeCl (64%) → 66 (CHO, H, SePh) + 67 (CHO, SePh, H) (9:1) (4.37)

Other enolic derivatives have also been used, such as compound **68**, which was transformed to the antibiotic marasmic acid (**69**) by Boeckman, as shown in Eq. 4.38.[104]

68 (H, H, COOMe, O) → PhSeBr, MeOH (92%) → (H, PhSe, OMe, O, H, COOMe, O) → DIBAL (95%) → (H, PhSe, OMe, O, H, COOMe, OH) → mCPBA (77%) → (H, CHO, CHO, H, COOMe) → BBr_3 → 69 (H, CHO, OH, O, H, O) (4.38)

4.1.3. Synthesis of α,β-unsaturated esters

The most efficient method for α-selenenylation of esters is the formation of their lithium enolates with lithium dialkyl amide and quenching with phenylselenenyl chloride, phenylselenenyl bromide or diphenyl diselenide.[5, 8, 11, 105] Sharpless exemplified the procedure in a synthesis of the methyl ester of the naturally occurring pheromone **70** (Eq. 4.39).[5] Since then, many other α,β-unsaturated esters have been prepared using this method. Several syntheses of natural products have benefited from this type of transformation. These include the synthesis of hinesol (**71**, Eq. 4.40),[106] the synthesis of pheromone **72** (Eq. 4.41),[107] Danishefsky's synthesis of the antibiotic pentalenolactone (**73**, Eq. 4.42),[108] and the synthesis of the alkaloid perhydrogephyrotoxin (**74**, Eq. 4.43).[109]

COOMe

(1) LDA–PhSeSePh

(2) $NaIO_4$ (80%)

COOMe

(4.39)

70

COOMe

(1) LDA–PhSeSePh

(2) $NaIO_4$ (90%)

COOMe

OH

H

(4.40)

71

CHO — (1) LiC≡CH; (2) $(EtO)_3CMe$, EtCOOH, Δ (95%) →

COOEt — (1) DIBAL; (2) CBr_4–PPh_3; (3) Mg, then CO_2; (4) MeOH, TsOH →

COOMe — (1) LDA–PhSeSePh; (2) $NaIO_4$ (85%) →

COOMe

72

(4.41)

H COOMe; O; O; H — (1) LDA–PhSeCl; (2) $NaIO_4$ (46%) → H COOMe; O; O; H

(1) DIBAL
(2) t-BuOOH, $VO(acac)_2$
(3) $CrO_3 \cdot H_2SO_4$
(4) KOH

H COOH; O; O; O; H

73

(4.42)

$H_{11}C_5$ H N H O O COOMe — (1) LDA–PhSeCl (2) $NaIO_4$ → $H_{11}C_5$ H N H O O COOMe

(1) Zn, AcOH
(2) NaOMe
(3) $LiAlH_4$

$H_{11}C_5$ H N H OH

74

(4.43)

With sterically hindered esters, the use of potassium hydride as the base seems to be more effective, as in Mander's synthesis of gibberellin A_4 (**75**, Eq. 4.44).[110]

H HO O OMe COOMe — (1) KH–PhSeSePh (2) H_2O_2 (74%) → H HO O OMe COOMe

O O H HO H COOH

75

(4.44)

Another base that has been used in this reaction is lithium 2,2,6,6-tetramethyl piperidide (LTMP). In Oppolzer's synthesis of *α*-kainic acid (**77**) two equivalents of the base had to be used in order to effect selenenylation of ester **76** (Eq. 4.45).[111]

COO^{t}Bu N OSi^{t}BuMe$_2$ EtOOC **76** —(1) LTMP (2 eq) (2) PhSeCl (3) H_2O_2, pyr (48%)→ COO^{t}Bu N OSi^{t}BuMe$_2$ EtOOC

—(1) Δ (ene reaction) (2) $F^{\ominus}$ (3) Jones oxidation (4) LiOH (5) CF_3COOH→ NH COOH COOH **77** (4.45)

Reduction of the ester function, either before or after the oxidation–*syn* elimination of the selenide, leads to primary allylic alcohols. This transformation was used in the synthesis of the pyrrolizidine alkaloids supinidine (**78**, Eq. 4.46)[112,113] and retronecine (**79**, Eq. 4.47).[114]

H COOEt N —LDA–PhSeCl (57%)→ H COOEt SePh N —$LiAlH_4$ (62%)→

H OH SePh N —H_2O_2 (59%)→ H OH N **78** (4.46)

PhCH$_2$O H COOMe — LDA–HMPA / PhSeSePh (95%) → PhCH$_2$O H COOMe SePh — mCPBA (90%) →

PhCH$_2$O H COOMe — (1) DIBAL / (2) Li–NH$_3$(l) (70%) → HO H OH **79** (4.47)

An alternative approach to retronecine (**79**), recently reported by Vedejs,[115] involves the selenenylation of aluminum enolate **81**, obtained by 1,4-addition of DIBAL to α,β-unsaturated ester **80** (Eq. 4.48).

o–NO$_2$C$_6$H$_4$CH$_2$O COOEt **80** — DIBAL → EtO OAl^{i}Bu$_2$ H **81** — PhSeCl (85%) → COOEt H SePh → **79** (4.48)

γ-Selenenylation of α,β-unsaturated esters is also possible. Using a variant of the reaction, Bryson has devised a convenient synthesis of substituted pyrroles, furans and thiophenes.[116] An example is shown in Eq. 4.49.

$$Cl(CH_2)_3C\equiv CCOOMe + H_2N(CH_2)_5COOMe \xrightarrow[(100\%)]{NaI,\ Na_2CO_3}$$

COOMe

N

COOMe

(1) LDA (1eq), PhSeBr

(2) H_2O_2

(71%)

COOMe

N

COOMe

(4.49)

4.1.4. Synthesis of α,β-unsaturated lactones

α-Selenenylation of lactones is performed in the same way as with esters, *i.e.* treatment with LDA and quenching with phenylselenenyl chloride, phenylselenenyl bromide or diphenyl diselenide.[5, 11] Depending on the substituents in the α- and β-position of the lactone, oxidative elimination of the phenylseleno group can produce an endocyclic olefinic bond (Eq. 4.50) or an exocyclic bond (Eq. 4.51), the former usually being the favored process. This selectivity is analogous to that observed for similar sulfoxide eliminations; according to Trost, it can be attributed to the relative stability of the conformations required for the elimination.[117] Thus, **82** is a more stable conformation than **84** because in **82** the dipole moments of the carbonyl and Se=O bonds are more closely aligned, leading to smaller dipole–dipole repulsion.

O SePh O H [O] O Ph Se O O H −PhSeOH O O

82 83

(4.50)

[O]

84

−PhSeOH

85

(4.51)

α,β-Unsaturated lactones (**83**) and particularly the 5-membered ring ones (butenolides)[118] are valuable synthetic intermediates. As shown from the examples below, they have been used for the functionalization of the α- or the β-position of the lactone, or both.

In a synthesis of the pseudoguaianolide confertin (**88**, Eq. 4.52),[119] hydrogenation of butenolide **87**, formed from **86**, led to an overall epimerization of the β-position.

(1) LDA–PhSeSePh
(2) H_2O_2

86

87

H_2, Pd/C

(4.52)

88

Formation of butenolide **90** from **89** (Eq. 4.53), followed by reduction to the diol and ozonolysis of the olefinic bond resulted in the conversion of **89** to α-hydroxy ketone **91**.[120] In this selenenylation, the base used was lithium isopropylcyclohexyl amide (LICA).

(1) LICA–PhSeBr
(2) AcOOH
(1) $LiAlH_4$
(2) O_3

89 **90** **91**

(4.53)

Transformation of lactone **92** to its α,β-unsaturated derivative (**93**), epoxidation of the olefinic bond of **93**, and reduction of the resulting epoxide produced, via an overall β-hydroxylation, the naturally occurring macrolide, diplodialide C (**94**, Eq. 4.54).[121]

(1) LDA–PhSeBr
(2) H_2O_2
(1) mCPBA
(2) Li, NH_3 (*l*), *t*-BuOH

92 **93** **94**

(4.54)

Conjugate addition to butenolide **96**, prepared from the optically active butanolide **95** (Eq. 4.55), and subsequent alkylation of the resulting lactone enolate, gave the all-*trans* α,β-dialkylated product (**97**), which was used in the synthesis of the antileukemic lignan isostegane (**98**) and several related compounds.[122]

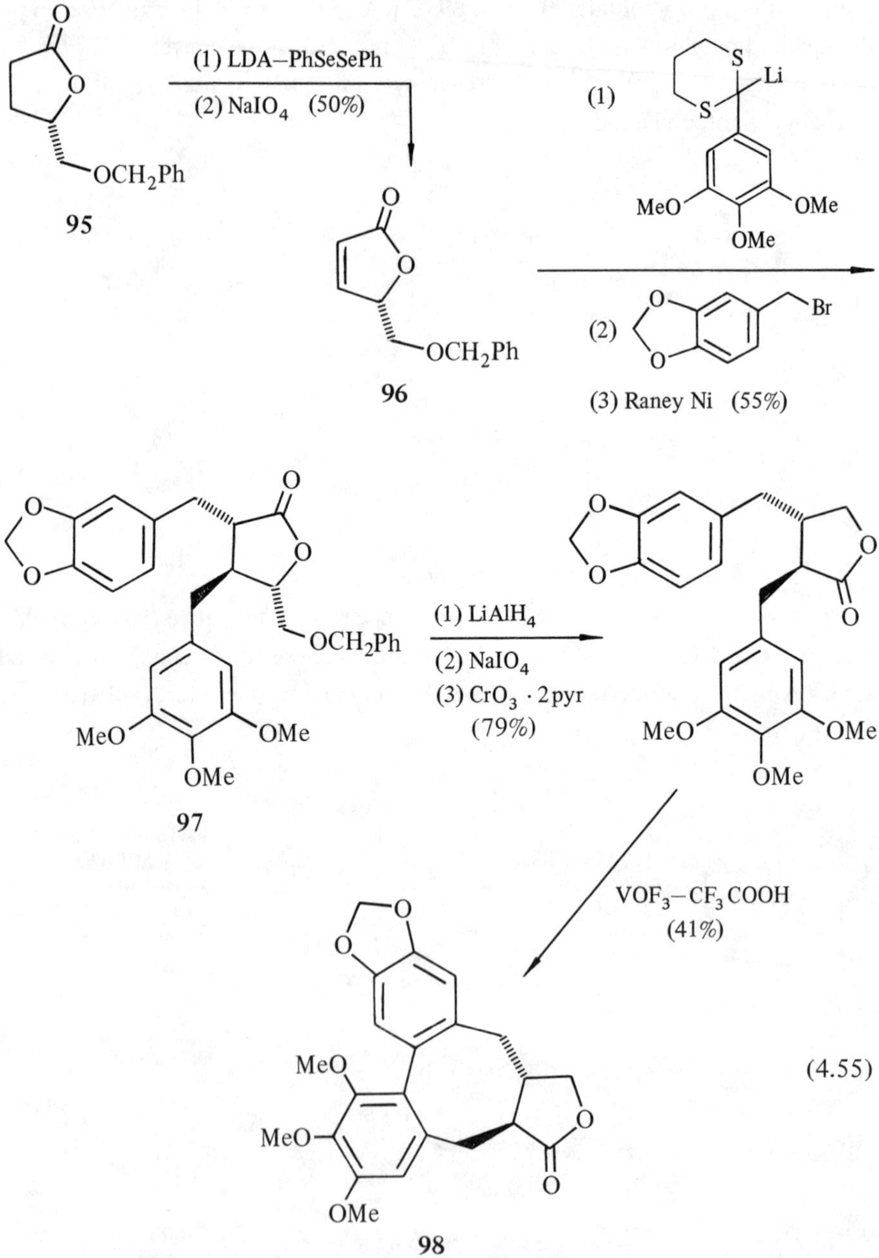

(1) LDA—PhSeSePh
(2) $NaIO_4$ (50%)

95 → 96

(1) 2-lithio-2-(3,4,5-trimethoxyphenyl)-1,3-dithiane
(2) 3,4-methylenedioxybenzyl bromide
(3) Raney Ni (55%)

97

(1) $LiAlH_4$
(2) $NaIO_4$
(3) $CrO_3 \cdot 2\,pyr$
(79%)

VOF_3—CF_3COOH
(41%)

98

(4.55)

Another example of conjugate addition to a butenolide leading to a trans β,γ-substituted butanolide is shown in Eq. 4.56.[123] The final product of this sequence **(99)** is the naturally occurring terpenoid pheromone eldanolide.[123]

(1) LDA–PhSeBr

(2) H_2O_2 (60%)

LiCuMe$_2$ (66%)

(4.56)

99

Photochemical cycloaddition of acetylene to lactone **100** was used in a short and direct route to intermediate **101**, which bears the tricyclic nucleus of verrucarol (**102**, Eq. 4.57).[124]

MeOOC

(1) $CH_2{=}C(OEt)CH{=}CH_2$

(2) $(CH_2OH)_2$, TsOH

(3) LiCuMe$_2$

(58%)

MeOOC

(1) LDA–PhSeCl

(2) H_2O_2 (49%)

(4.57)

HC≡CH, $h\nu$

(62%)

MeOOC

MeOOC

100

(1) DIBAL

(2) Ac_2O, pyr

(3) TsOH, AcOH H_2O–C_6H_6

(4) Ac_2O, pyr

OAc

MeOOC

HO

OH

101

102

The facile formation of butenolides of type **105** from the corresponding selenoxides (**104**), combined with DIBAL reduction of the butenolides to the furans (**106**) is the basis for a general and efficient synthesis of substituted furans developed by Grieco (Eq. 4.58).[125]

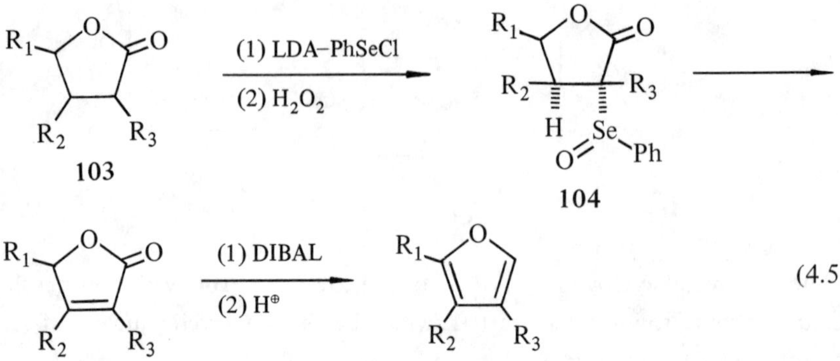

(1) LDA–PhSeCl
(2) H_2O_2

103 **104**

(1) DIBAL
(2) $H^{\oplus}$

105 **106** (4.58)

α,β-Unsaturated lactones of the general types **83** (Eq. 4.50) and **85** (Eq. 4.51) occur in many natural products. As demonstrated below, selenium methodology has been very helpful in their construction.

The olefinic bond in **108**, an intermediate in Semmelhack's synthesis of fomannosin (**109**, Eq. 4.59), was nicely introduced by selenoxide elimination.[126] However, the selenenylation of the dianion of **107** could not be made to go to completion, and 40–45% of the starting material was recovered.

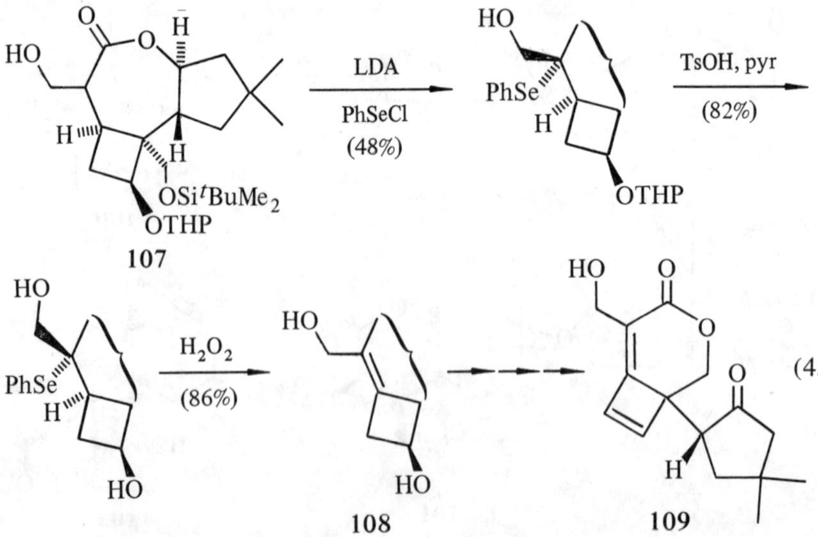

LDA
PhSeCl
(48%)

TsOH, pyr
(82%)

107

H_2O_2
(86%)

108 **109** (4.59)

The selenenylation of intermediate **110**, used in Smith's synthesis of paniculide A (**114**, Eq. 4.60) was also difficult.[127, 128] The reaction failed completely with LDA as the base. However, by using lithium 2,2,6,6-tetramethylpiperidide (LTMP) under strictly defined conditions, the desired selenide (**112**) could be obtained, but only in moderate yield.[127] Nevertheless, the conditions used were not suitable for the selenenylation of the more highly oxygenated intermediate **111** needed for the synthesis of paniculide B (**115**, Eq. 4.60),[128] probably due to strong coordination of lithium with the oxygens of the molecule. In the end, the conversion of **110** to **112** and the conversion of **111** to **113** were achieved by using a more highly dissociated base, namely potassium bis-(trimethylsilyl) amide, and diphenyl diselenide as the source of electrophilic PhSe. It was also necessary to oxidize the by-product potassium phenylselenolate by passing oxygen into the reaction mixture prior to workup. Otherwise, the selenide formed (**112** or **113**) was deselenated *in situ* by the potassium phenylselenolate. In another approach to paniculide A (**114**), the selenenylation step (**116**→**117**) could be effected by LDA and phenylselenenyl chloride in the presence of hexamethylphosphoric triamide (HMPA), as shown in Eq. 4.61.[129]

$(Me_3Si)_2NK$
PhSeSePh, O_2

110: R = H
111: R = OSi^{t}BuMe$_2$

112: R = H (82%)
113: R = OSi^{t}BuMe$_2$ (72%)

(1) $NaIO_4$
(2) AcOH, THF, H_2O (81%)

114: R = H
115: R = OH

(4.60)

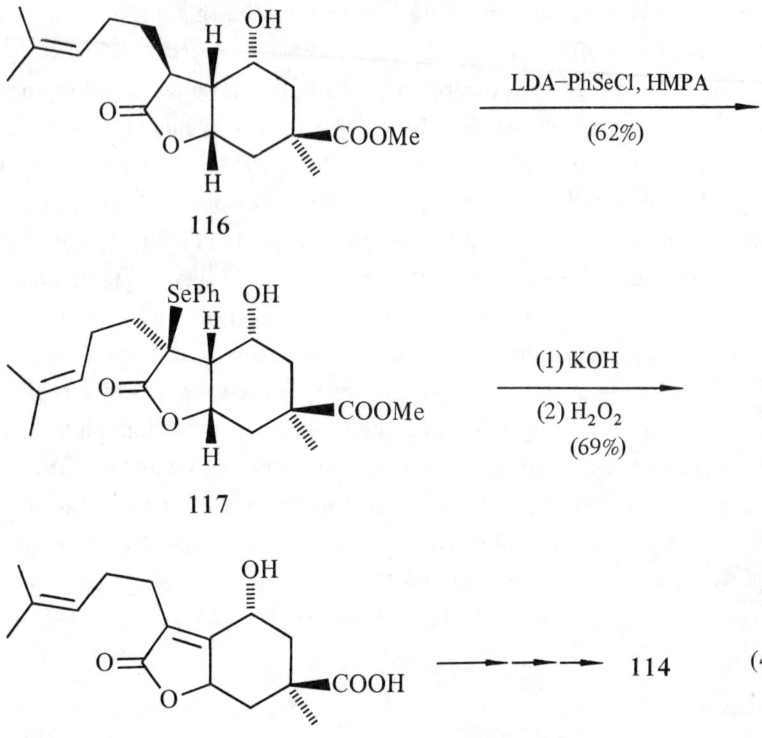

LDA–PhSeCl, HMPA (62%)

116

(1) KOH
(2) H_2O_2 (69%)

117

→ → → 114 (4.61)

Selenenylation of **118**, followed by oxidative removal of the PhSe group, gave **119** (Eq. 4.62),[130] a model of the clerodane diterpenes such as olearin (**120**). A similar sequence was used in the final stages of a synthesis of the anticancer agent aplysistatin (**121**), outlined in Eq. 4.63.[131]

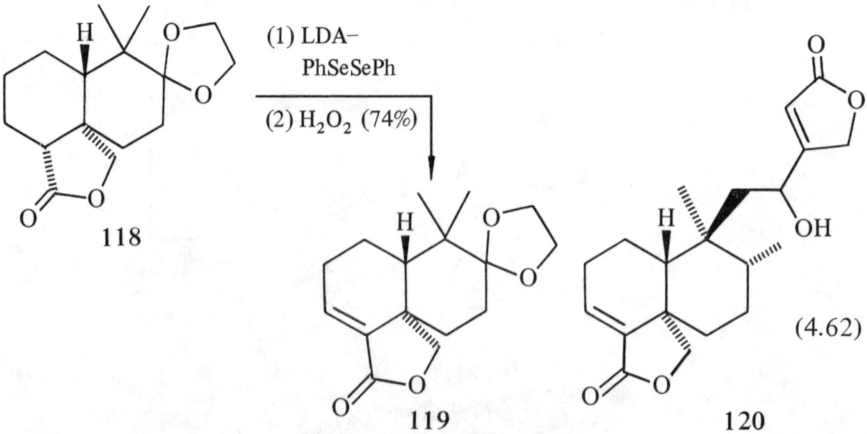

(1) LDA–PhSeSePh
(2) H_2O_2 (74%)

118 119 120 (4.62)

(4.63)

α-Methylene lactones (**85**, $R^1 = R^2 = H$) and especially the 5-membered ring ones, are found in the structures of a variety of natural products, most of which are biologically active. Because of this, a lot of synthetic studies were directed towards this structural unit.[132] Grieco has introduced the use of organo-selenium methodology in this area with remarkable success.[133] Grieco's method relies on the conversion of the α-unsubstituted lactones, such as **122** (Eq. 4.64) and **125** (Eq. 4.65), to the corresponding α-methyl-α-phenylseleno lactones (**123** and **126**), in which the PhSe group is *trans* to the β-hydrogen. Both the *trans*-fused lactones, such as **124**, and the *cis*-fused lactones, such as **127**, can be prepared in high yield by choosing the right methylation–selenenylation sequence.

(4.64)

LDA–PhSeSePh

LDA–MeI

125

H_2O_2

126

127

(4.65)

Several syntheses of sesquiterpene α-methylene γ-lactones have used this methylenation approach. These include the syntheses of tuberiferine (**128**, Eq. 4.66[134] and Eq. 4.67[135]), artecalin (**129**, Eq. 4.68),[134] arglanine (**130**, Eq. 4.69)[136] and costunolide (**131**, Eq. 4.70).[137] Similar reactions were also used in the synthesis of zaluzanin C and zaluzanin D.[230]

LDA–PhSeSePh (66%)

Jones oxidation (56%)

MnO_2 (87%)

H_2O_2 (100%)

128

(4.66)

(1) LDA – PhSeSePh
(2) HCl aq
(85%)

SePh

O

H

O

O

LDA–PhSeCl
(76%)

PhSe

O_3
(60%)

SePh

128

(4.67)

LDA–PhSeSePh
(81%)

HO

SePh

MnO_2 (60%)

OH

129

SePh

Zn, HOAc
(95%)

H_2O_2
(75%)

(4.68)

OH OH LDA–PhSeSePh SePh H H O O O O

(1) MnO_2
(2) $(CH_2OH)_2$ TsOH
(3) $H_3O^{\oplus}$, O_2
(4) H_2O_2

O HO H O O

130

(4.69)

(1) LDA–PhSeSePh
(2) H_2O_2

H O H O O

Δ

O O

131

(4.70)

Reduction of α-phenylselenolactones to β-hydroxy selenides, followed by formation of the selenoxides and *syn* elimination, leads to allylic alcohols. The sequence was used in a recent synthesis of retronecine (**79**, Eq. 4.71).[138]

LDA–PhSeSePh / THF, HMPA (21%) ; $LiAlH_4$ (95%) ; H_2O_2 (53%)

79

(4.71)

4.1.5. Synthesis of α,β-unsaturated lactams

There are only a few examples of olefination of lactams by organoselenium methods. However, the α,β-unsaturated lactam functionality is not very abundant in natural products. In general, selenenylation and the oxidation of the selenide can be carried out similarly to other carbonyl compounds, but in order to avoid diselenated products, two equivalents of base must be used.[139] An example is shown in Eq. 4.72.[139]

LDA (2 eq) / PhSeCl (55%) ; H_2O_2 (57%)

(4.72)

The selenenylation of the strained lactam **132** was performed without the use of base and gave the chlorine-containing adduct **133** (Eq. 4.73).[140] The product was converted to the alkaloid erysotramidine (**134**) via oxidative elimination of the PhSe group, substitution of chloride by methoxyl, and desulfurization (Eq. 4.73).

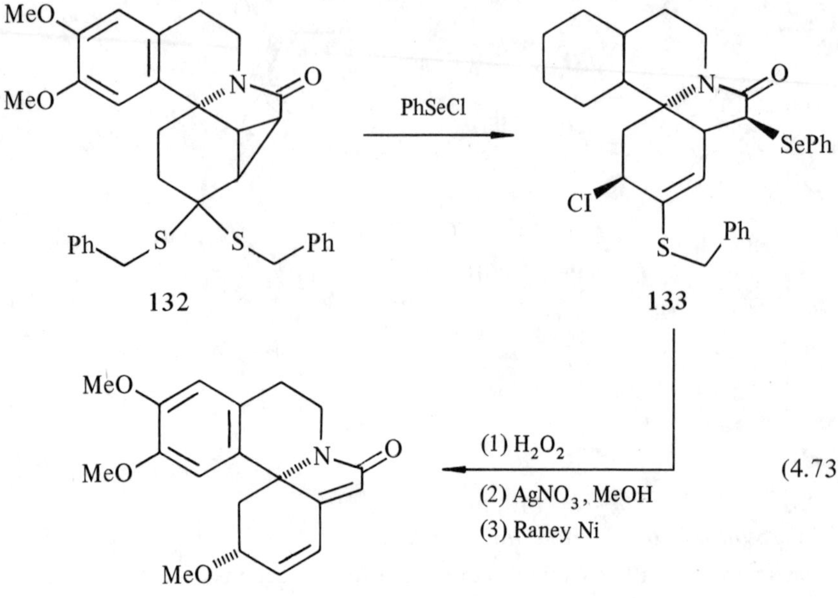

MeO MeO N O S S Ph Ph **132** PhSeCl N O SePh Cl S Ph **133** (1) H_2O_2 (2) $AgNO_3$, MeOH (3) Raney Ni MeO MeO N O MeO **134** (4.73)

An example of olefination of a substituted succinimide is found in the synthesis of the antitumor antibiotic showdomycin (**135**, Eq. 4.74).[141]

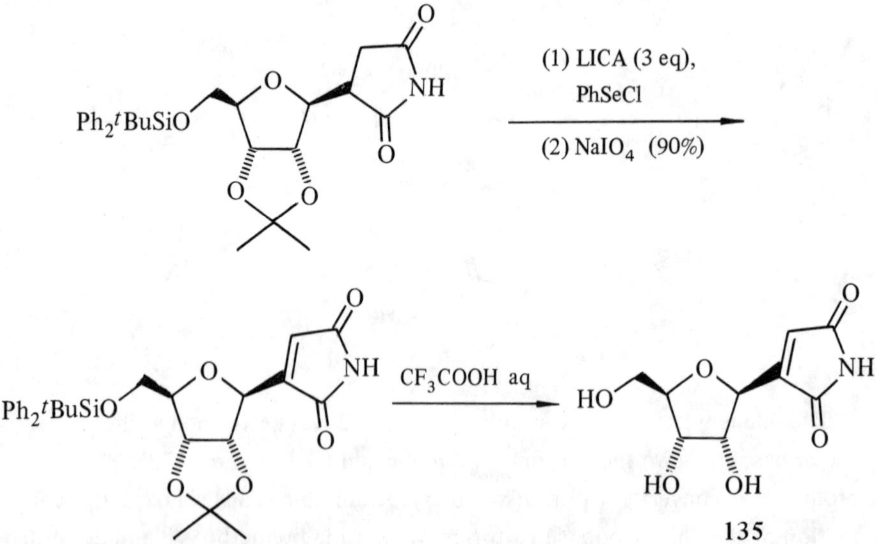

Ph$_2$^{t}BuSiO O O O NH O O (1) LICA (3 eq), PhSeCl (2) $NaIO_4$ (90%) Ph$_2$^{t}BuSiO O O O NH O O CF_3COOH aq HO O HO OH NH O O **135** (4.74)

Another antibiotic that posseses the exocyclic α,β-unsaturated lactam functionality is mycelianamide (**136**), which was synthesized by Cava[142] as indicated in Eq. 4.75.

CH NO$_2$ COOEt

(1) PhSeH, pyr
(2) Al-Hg (50%)

NHOH COOEt SePh

(1) MeCHBrCOCl, i-Pr_2NEt
(2) NH_2OH
(3) $Me_2{}^tBuSiCl$, imidazole

$^tBuMe_2Si$ O O N Me N OSi^tBuMe_2 SePh O

(1) mCPBA
(2) HF, EtOH

HO N O Me N OH H O

136

(4.75)

4.1.6. *Synthesis of α,β-unsaturated nitriles*

The α-selenenylation of nitriles was studied by Heathcock.[143] The best results are obtained by treatment of the saturated nitrile with 2 equivalents of lithium isopropylcyclohexyl amide (LICA) and quenching the resulting anion with diphenyl diselenide or with phenylselenenyl bromide. The method was applied by Heathcock in an approach to the skeleton of dendrobine (**140**) and it was used as a means of epimerizing the β-carbon of **137** to give **139** via **138**, as shown in Eq. 4.76.[144]

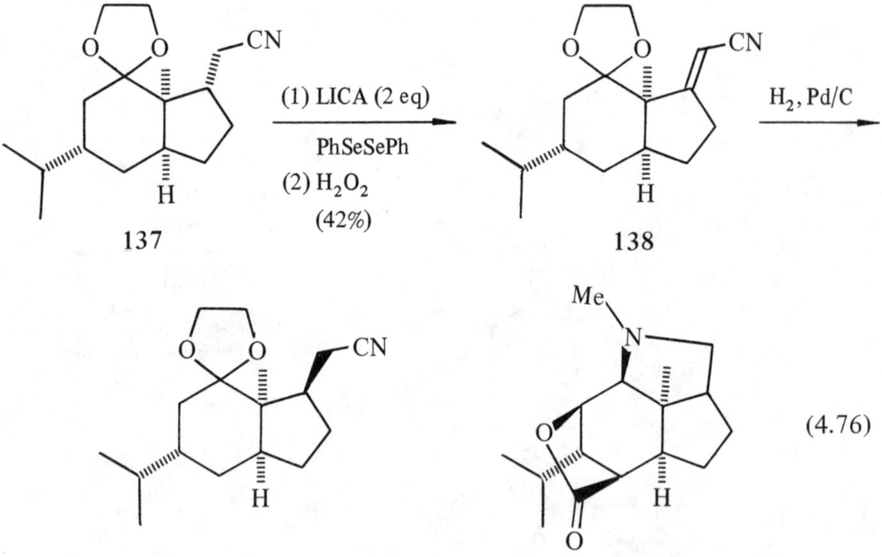

(1) LICA (2 eq), PhSeSePh; (2) H_2O_2 (42%)

H_2, Pd/C

137 **138** **139** **140** (4.76)

An alternative synthesis of α,β-unsaturated nitriles has been developed by Grieco.[145] It starts with an aldehyde (**141**), which is treated with *o*-nitrophenylselenocyanate (o-$O_2NC_6H_4SeCN$) in the presence of tri-*n*-butylphosphine, to give the homologated α-selenenylated nitrile (**142**), which upon oxidative elimination affords the α,β unsaturated nitrile (**143**, Eq. 4.77).[145]

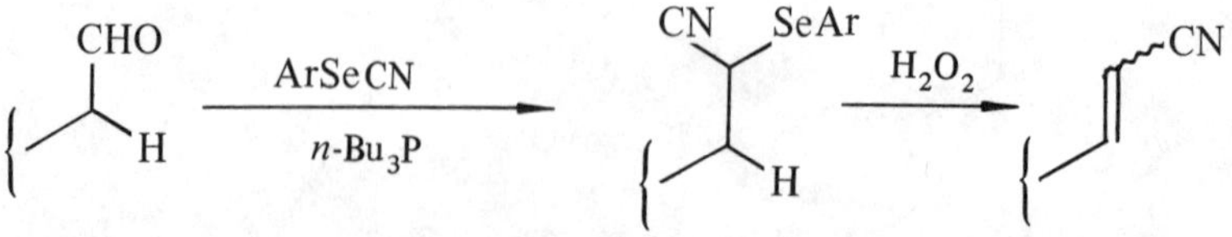

ArSeCN, *n*-Bu_3P

H_2O_2

141 **142** **143** (4.77)

4.1.7. Synthesis of α,β-unsaturated dicarbonyl compounds

In general, β-dicarbonyl compounds can be selenenylated similarly to the monocarbonyl derivatives, *i.e.* by reacting their enolates with phenylselenenyl chloride or bromide.[9,11,146] The most common base used is sodium hydride. For the oxidation of the selenides of five- and six-membered cyclic β-dicarbonyl compounds, hydrogen peroxide cannot be used, because it would result in epoxide formation and decomposition.[9,11] Ozonolysis is a more efficient oxidation method for such selenides. A typical example is given in Eq. 4.78.

NaH–PhSeCl (86%) → SePh → O_3 (90%) →

(4.78)

The five-membered ring β-ketoester **144**, prepared in similar fashion, as is shown in Eq. 4.79,[9,11,146,147] was used in a synthesis of the antibiotic sarcomycin (**145**).[147] This preparation of **144** is more efficient than the use of selenium dioxide, described in Eq. 3.14.

COOMe → (1) NaH–PhSeCl (2) O_3 (81%) → COOMe **144** → COOH **145**

(4.79)

An alternative procedure for the selenenylation of β-dicarbonyl compounds, involving the use of *N,N*-diethylbenzeneselenamide ($PhSeNEt_2$), prepared from phenylselenenyl chloride or bromide and diethyl amine, was also reported by Reich.[42] An example is shown in Eq. 4.80.

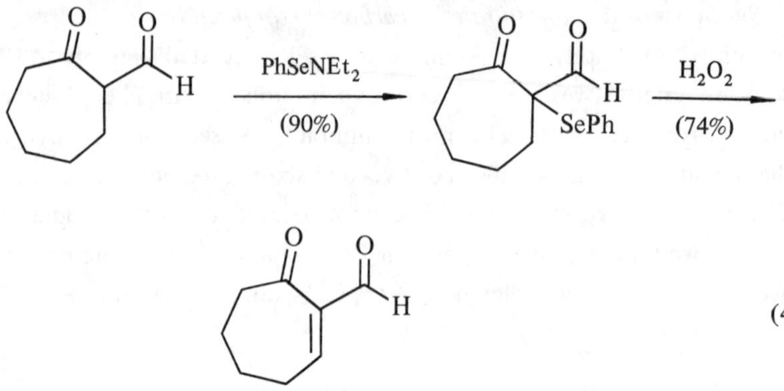

PhSeNEt$_2$ (90%)

H_2O_2 (74%)

(4.80)

In a synthesis of the antifungal mold metabolite LL-21271a (**147**), outlined in Eq. 4.81, selenenylation of hydroxymethylene ketone **146** was effected with phenylselenenyl chloride in the presence of triethylamine.[148]

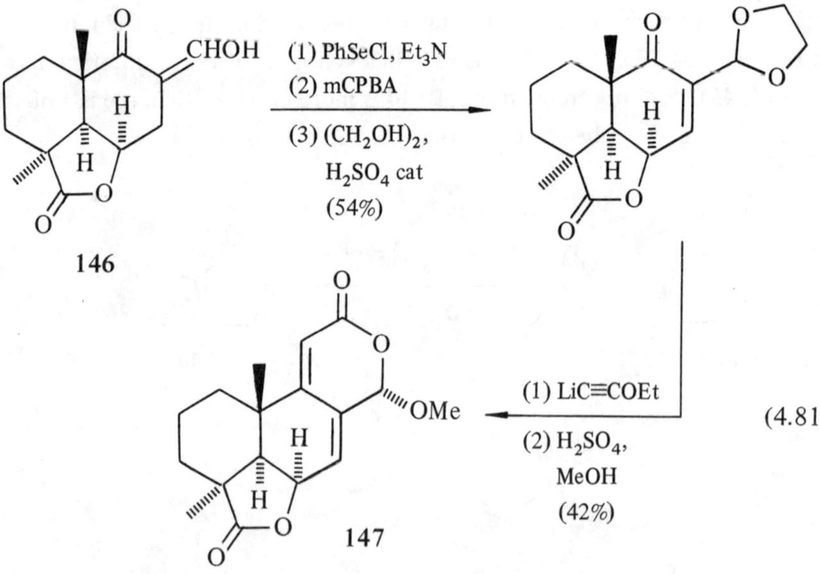

(1) PhSeCl, Et_3N
(2) mCPBA
(3) $(CH_2OH)_2$, H_2SO_4 cat
(54%)

146

(1) LiC≡COEt
(2) H_2SO_4, MeOH
(42%)

147

(4.81)

The use of phenylselenenyl chloride–pyridine for the selenenylation of β-dicarbonyl compounds was studied by Liotta.[149] The reagent was found to be very efficient and quite practical. Removal of the pyridine and oxidation *in situ* with hydrogen peroxide, gives the unsaturated product in high yield. Two typical examples are presented in Eq. 4.82[149] and Eq. 4.83.[149] The method was recently used in the synthesis of uridine derivatives (Eq. 4.83a).[225]

(1) PhSeCl–pyr
(2) H_2O_2
(95%)

(4.82)

(1) PhSeCl–pyr
(2) H_2O_2 (85%)

(4.83)

(1) PhSeCl–pyr
(2) H_2O_2
(3) CF_3COOH aq

(4.83a)

In some cases, when excess of phenylselenenyl chloride–pyridine is used, the unsaturated product is obtained directly via a nonoxidative elimination of the PhSe group.[149] Such was the case in a synthesis of waburganol (**45**), described earlier in Eq. 4.27.[95]

β-Dicarbonyl compounds can also be converted to their α,β-unsaturated derivatives by treating their enolates, obtained with sodium hydride in THF–HMPA, with elemental selenium (Se), quenching the selenolate formed with methyl iodide and oxidatively eliminating the resulting MeSe group.[150] The procedure is simple, effective and relatively inexpensive, and so it can be applied to large-scale preparations. It is applicable to β-diketones, β-diesters and β-ketolactones, but not to β-ketoaldehydes. Examples are given in Eq. 4.84,[150] Eq. 4.85,[150] and Eq. 4,86.[150]

(1) NaH, THF–HMPA
(2) Se
(3) MeI
(82%)

SeMe

H_2O_2
(98%)

(4.84)

(1) NaH, THF–HMPA
(2) Se
(3) MeI
(84%)

SeMe

H_2O_2
(96%)

(4.85)

(1) NaH, THF–HMPA
(2) Se
(3) MeI
(78%)

EtO OEt SeMe

H_2O_2
(96%)

(4.86)

γ-Selenenylation of β-dicarbonyl compounds is possible through their dianions, formed with LDA in the presence of HMPA, as in a synthesis of diplodialide A (**148**), shown in Eq. 4.87.[151]

LDA (2 eq),
THF–HMPA
PhSeBr
(38%)

SePh

$NaIO_4$
(30%)

(4.87)

148

4.2 Functionalization of olefins

Electrophilic organoselenium reagents (i. e. PhSeX) add readily to olefinic bonds (Eq. 4.88). Presumably, the selenonium intermediate **150** is formed, which suffers nucleophilic attack from $X^{\ominus}$ to give the *trans*-substituted selenide **151**. Table 4.3 shows the various PhSeX reagents that were found to add to olefins. (See Table 4.1 for the corresponding conditions.)

PhSeX SePh $X^{\ominus}$ SePh X (4.88)

149 **150** **151**

Table 4.3. Electrophilic PhSeX reagents that add to olefins

X (of PhSeX)	*Ref.*
$OCOCF_3$	10, 25, 26
OAc	6, 42
OR	6, 27, 28
OH	28, 33–36, 41, 45
CN	32
SO_2Ar	46–48, 49
Cl	6, 18,19
Br	6, 21–23
NHCOMe	44
NO_2	45

Depending on the nature of X, oxidation–syn elimination of the β-substituted selenides (**151**) proceeds away from the X group, giving the allylic substituted product (**152**), or towards the X group, giving the X-substituted olefin (**153**), as indicated in Eq. 4.89. Contrary to the general rule, however, the non-availability of an α-hydrogen in the preferred direction, may result in an "abnormal" elimination product.[21, 155, 158] Also, in some cases (when X = Cl, Br or N_3) both types of product are obtained.

151 → (1) [O] (2) Δ → 152 ; 151 → (1) [O] (2) Δ → 153 (4.89)

Table 4.4. Oxidative elimination of β-substituted selenides.

151→152		151→153	
X	*Ref.*	*X*	*Ref.*
OAc	6	Cl	6, 18, 19
OR	6, 27, 28	Br	6, 21–23
OH	6, 33–36, 41, 45	N_3	23
NHCOMe	44	CN	32
Cl	6, 18, 19	SO_2Ar	46–48, 49
Br	6, 21–23	NO_2	45
N_3	23		

One of the most useful reactions of the type shown in Eq. 4.88 and Eq. 4.89, is the alkoxyselenation reaction, *i.e.*, addition of the elements of PhSeOR to olefins.[6,27,28] The reaction can be applied in the preparation of an allyl alkyl ether functionality, as in the synthesis of the alkaloid crinamine (**154**), outlined in Eq. 4.90.[152]

(1) PhSeSePh, MeCONHBr, MeOH; (2) H_2O_2 (35%)

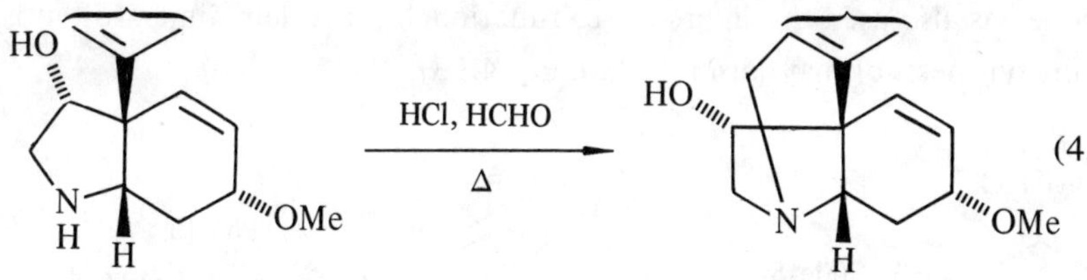

HCl, HCHO
Δ

(4.90)

154

Application of the alkoxyselenation reaction to dihydropyrans (**155**) gives addition products (**156**) regiospecifically in which the alkoxy group is next to the oxygen and the PhSe group is away from it, as shown in Eq. 4.91.[27] The products of oxidative elimination of the PhSe group (**157**) are useful synthetic intermediates. With substituted dihydropyrans, the reaction also proceeds with some stereoselectivity.

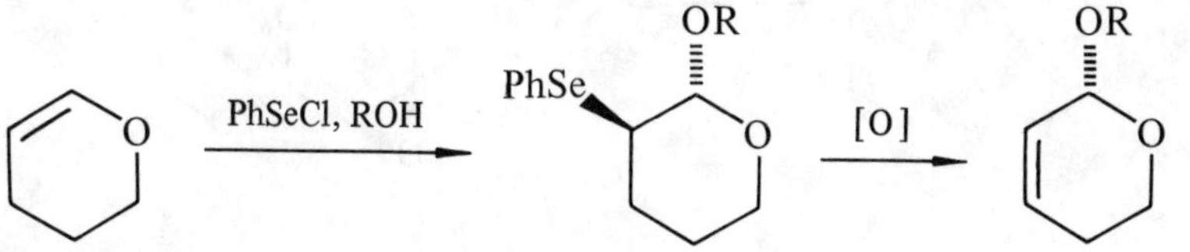

PhSeCl, ROH
[O]

(4.91)

155 156 157

This methodology was used by Kozikowski in his synthesis of the antibiotic pseudomonic acid C (**158**, Eq. 4.92).[153]

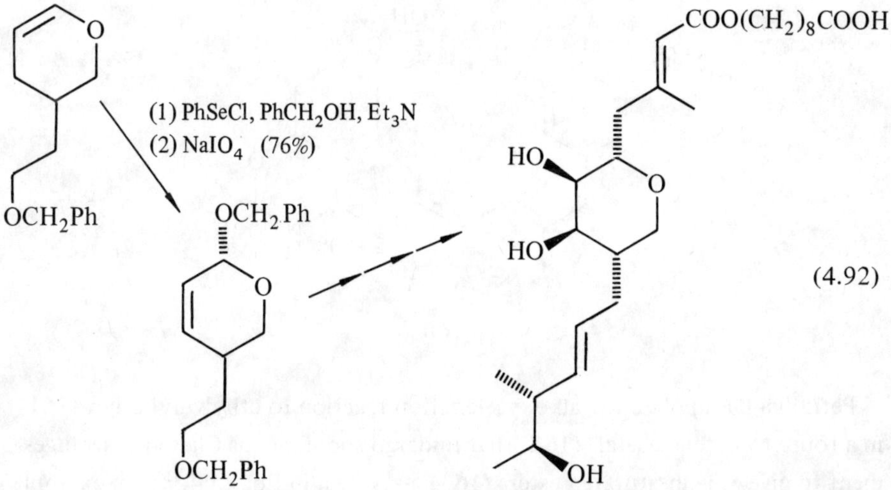

(1) PhSeCl, $PhCH_2OH$, Et_3N
(2) $NaIO_4$ (76%)

$COO(CH_2)_8COOH$

(4.92)

158

Isobe has also used this approach to functionalize acrolein dimer **159** in his elegant synthesis of maytansinol (**160**, Eq. 4.93).[154]

CHO

O

159

$PhS(Me_3Si)_2CLi$

SPh

Me_3Si

O

(1) PhSeCl, $HO(CH_2)_2OMe$

(2) mCPBA

SO_2Ph

Me_3Si

O O O

mCPBA

SO_2Ph

Me_3Si

O O O O

Cl O OH

MeO N O H

O

N O

HO H

MeO

160

(4.93)

Petrzilka has applied the alkoxyselenation reaction to ethylvinyl ether (**161**) in a route to ketene acetals (**162**) that undergo spontaneous Claisen rearrangement to give γ,δ-unsaturated esters (**163**), as shown in Eq. 4.94.[155] An example is given in Eq. 4.95.

$CH_2{=}CHOEt$ **161** —PhSeBr, HO–C(R_1)(R_2)–C(R_3)=C(R_5)R_4→ (1) $NaIO_4$, (2) Δ

162 → **163** (4.94)

PhSeBr, $CH_2{=}CHOEt$; (1) $NaIO_4$, (2) Δ (95%)

COOEt (4.95)

This novel Claisen-type transformation was recently employed in the conversion of **164** to **165** (Eq. 4.96),[156] an intermediate in the synthesis of guaianolides.

164 —(1) PhSeBr, $CH_2{=}CHOEt$; (2) $NaIO_4$; (3) Δ (45%)→ **165** (4.96)

Addition of the elements of PhSeOAc to an olefin was used in a synthesis of compactin (**168**, Eq. 4.97), as part of an overall transformation of olefin **166** to enone **167**.[157]

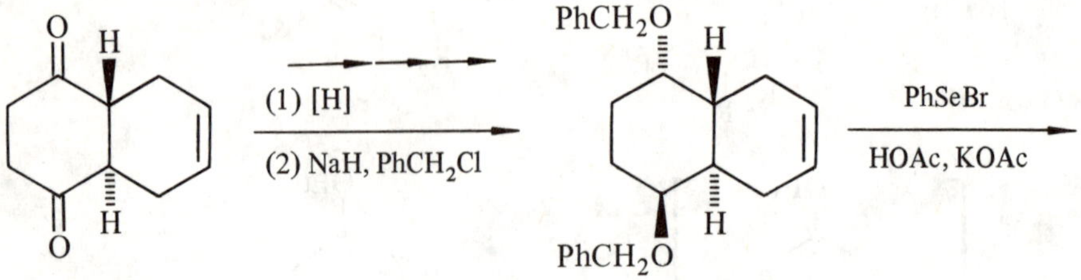

(1) [H]

(2) NaH, $PhCH_2Cl$

$PhCH_2O$ H H $PhCH_2O$

166

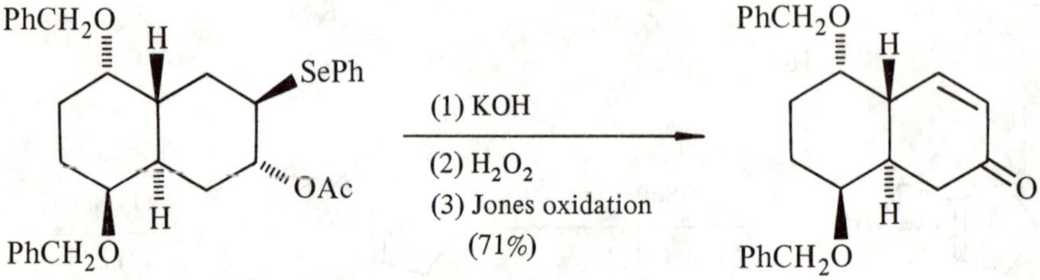

PhSeBr

HOAc, KOAc

$PhCH_2O$ H SePh H OAc $PhCH_2O$

(1) KOH

(2) H_2O_2

(3) Jones oxidation

(71%)

$PhCH_2O$ H H O $PhCH_2O$

167

HO O O O O H

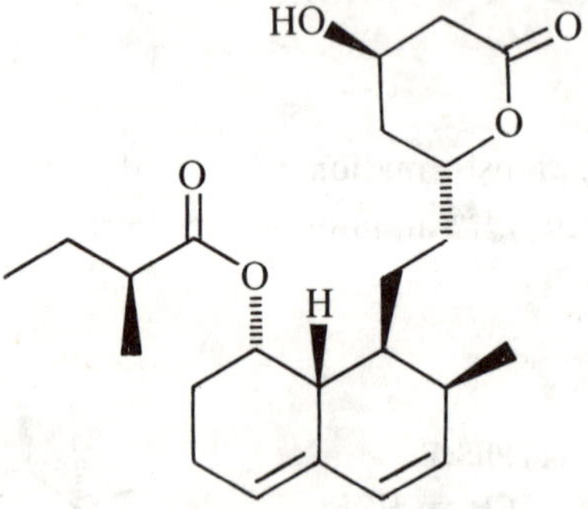

168

(4.97)

Regiospecific addition of PhSeOAc to steroidal olefin **169** gave adduct **170**, which upon oxidative elimination afforded "abnormal" product **171** (Eq. 4.98).[158] The alternative, "normal" elimination pathway is not possible in this case, since the β-hydrogen is *trans* to the PhSe group.

OAc H H H AcO 169 PhSeBr, KOAc (80%)

OAc H H H AcO SePh AcO 170 H_2O_2 (90%) OAc H H H AcO AcO 171 (4.98)

Formation of β-hydroxyselenides (Eq. 4.88, X = OH) by addition of the elements of PhSeOH to olefins, has been achieved by several methods. These include: the use of $PhSeOCOCF_3$,[10, 26] or PhSeOAc,[6] followed by hydrolysis; the use of PhSeOH generated *in situ* by comproportionation of PhSeSePh and $PhSeO_2H$;[33, 34] or by reduction of $PhSeO_2H$ with H_3PO_2;[35] the cuprous chloride-catalyzed reaction of PhSeCN in water;[28] the reaction of PhSeCl in aqueous medium;[36] and the use of *N*-phenylselenosuccinimide (N-PSS) or *N*-phenylselenophthalimide (N-PSP) in the presence of water and an acid catalyst.[45] These reactions proceed with considerable regioselectivity, as shown in the examples presented in Eq. 4.99,[10] 4.100,[33] 4.101[36] and 4.102.[45] The favored product is usually the Markownikov adduct, *i.e.*, the product formed by the addition of PhSe to the less substituted carbon.

$PhSeOCOCF_3$ (69%) $OCOCF_3$ SePh KOH OH SePh

(4.99)

"PhSeOH," $MgSO_4$ (4.100)

PhSeCl, H_2O (73%) (4.101)

N-PSS or N-PSP, H_2O (84–89%) (4.102)

Oxidative elimination of the PhSe group of β-hydroxyselenides (**151**, X = OH) results in an overall conversion of olefins (**149**) to allylic alcohols (**152**, X = OH). An application is shown in the synthesis of a coformycin analog (**172**, Eq. 4.103).[159]

N-PSP, H_2O (47%); mCPBA; NH_3, MeOH (80%) (4.103)

172

Addition of the elements of PhSeCl to olefins proceeds initially in the anti-Markownikov fashion, while longer reaction times and higher temperatures give the thermodynamically more stable Markownikov adducts, *i.e.*, with the PhSe group attached to the less substituted carbon.[18, 22] Thus, a terminal olefin can be converted to an allylic chloride (derived from the anti-Markownikov adduct) only if the addition of PhSeCl is done at low temperature. An example of this reaction is found in a recent approach to the antibiotic pseudomonic acid C (**158**, Eq. 4.92), as shown in Eq. 4.104.[160]

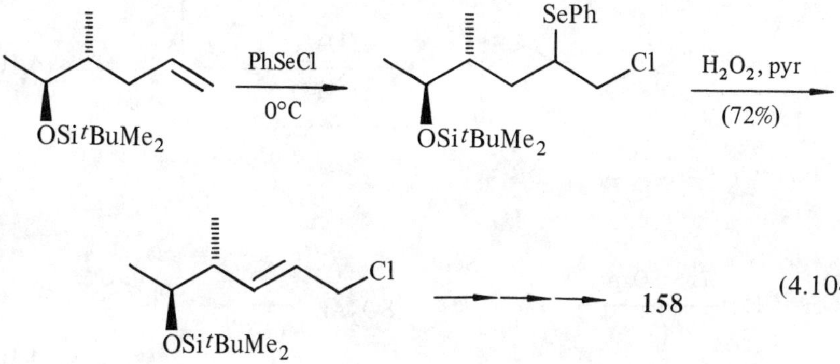

OSi^{t}BuMe$_2$ —PhSeCl, 0°C→ SePh, Cl, OSi^{t}BuMe$_2$ —H_2O_2, pyr (72%)→

Cl, OSi^{t}BuMe$_2$ ⇢ **158** (4.104)

Liotta has studied in detail the reaction of phenylselenenyl chloride with allylic alcohols and their acetates and has found that the addition proceeds with high regiocontrol (and sometimes high stereocontrol).[19] The regioselectivity was nicely applied in a novel 1,3-enone transposition, exemplified in Eq. 4.105.[19]

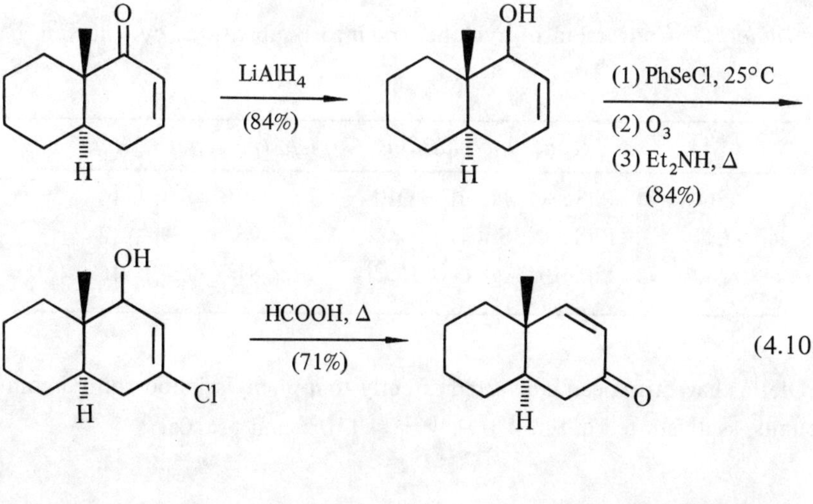

O, H —$LiAlH_4$ (84%)→ OH, H —(1) PhSeCl, 25°C; (2) O_3; (3) Et_2NH, Δ (84%)→

OH, Cl, H —HCOOH, Δ (71%)→ O, H (4.105)

The overall functionalization of olefins (**149**) to X-substituted olefins (**153**) has been used to prepare vinyl bromides,[21] vinyl sulfones,[46, 47] α,β-unsaturated nitriles[32] and nitroolefins. Typical examples are shown in Eq. 4.106,[21] 4.107[47] and 4.108.

PheSeBr / MeCN → SePh, Br; (1) O_3 (2) i-Pr_2NH, Δ (85%) → Br (4.106)

$PhCH{=}CH_2$ —$PhSeSO_2Ar$, $h\nu$ (73%)→ SePh, Ph, SO_2Ar —[O] (77%)→ Ph, SO_2Ar (4.107)

(See Table 4.5) → SePh, X —H_2O_2 (93%)→ X (4.108)

Table 4.5. Conversion of cyclohexene into 1-substituted cyclohexenes (Eq. 4.108).

X	*Reaction conditions*	*Yield (percent)*	*Ref.*
SO_2Ar	$PhSeSO_2Ar$, $BF_3 \cdot OEt_2$	89	46
CN	PhSeCN, $SnCl_4$	96	32
NO_2	PhSeBr, $AgNO_2-HgCl_2$	81	45

Olefins have also been converted directly to α-phenylselenocarbonyl compounds, as illustrated in Eq. 4.109,[37, 228] 4.110[37] and 4.110a.[228]

$$PhCH{=}CH_2 \xrightarrow[\text{or PhSeSePh–}Br_2\text{–}(Bu_3Sn)_2O\text{ (74\%)}]{\text{PhSeSePh–}t\text{-BuOOH (87\%)}} PhC(=O)CH_2SePh \quad (4.109)$$

(4.110)

PhSeSePh / PhSeOSePh (with O O); KF (71%)

(4.110a)

(1) PhSeSePh, Br_2, $(Bu_3Sn)_2O$ (73%)
(2) K_2CO_3 (76%)

Electrophilic organoselenium reagents also add to enol ethers, silyl enol ethers and enol acetates, leading to α-phenylseleno carbonyl compounds. Some examples are discussed in §4.1.

Nucleophilic selenium reagents add to olefins conjugated with carbonyl via a Michael-type reaction. Examples are given in §4.4 and §8.2.

4.3 Conversion of epoxides to allylic alcohols

Epoxides (**173**) are readily cleaved by nucleophilic organoselenium reagents, PhSeM, (Table 4.1), giving β-hydroxyselenides (**174**), which are transformed to allylic alcohols (**175**) by oxidation and *syn* elimination (Eq. 4.111). This mild procedure for the conversion of epoxides (**173**) to allylic alcohols (**175**) was developed by Sharpless.[3] An important feature of the sequence is that the opening of the epoxide ring proceeds almost regiospecifically, attaching the PhSe group at the less substituted carbon. For example, myrcene monoepoxide (**176**) gave exclusively the naturally occurring isomyrcenol (**177**, Eq. 4.112).[3]

173 —PhSeM→ **174** —(1) [O], (2) Δ→ **175**

(4.111)

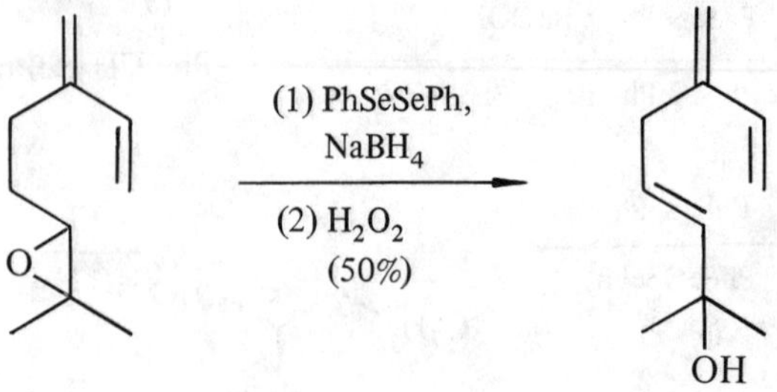

(1) PhSeSePh, $NaBH_4$

(2) H_2O_2 (50%)

OH

176 177

(4.112)

The method has found wide application in natural products synthesis. It was used in the synthesis of lycoricidine (**178**, Eq. 4.113),[161] lycorine (**179**, Eq. 4.114),[162] senepoxide (**180**, Eq. 4.115),[163] chorismic acid (**181**, Eq. 4.116)[164] and other compounds.

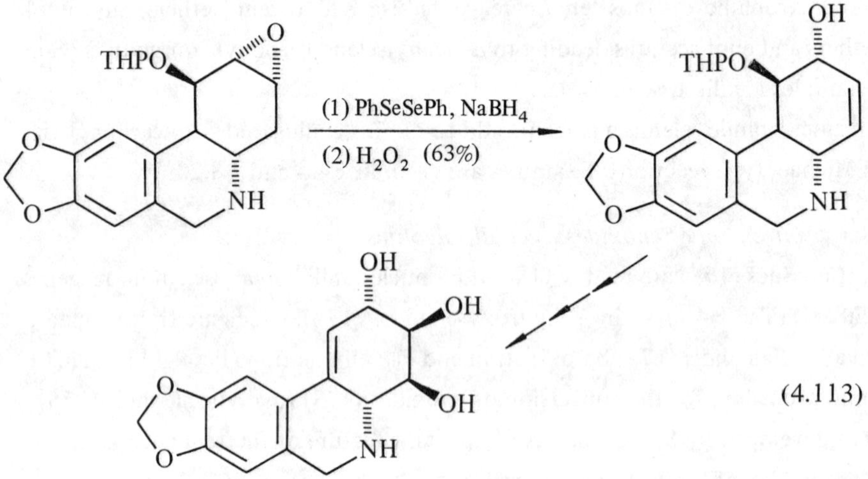

THPO O

O O NH

(1) PhSeSePh, $NaBH_4$

(2) H_2O_2 (63%)

OH THPO

O O NH

OH OH OH

O O NH

178

(4.113)

O H H O O H N O

(1) PhSeSePh, $NaBH_4$

(2) $NaIO_4$ (70%)

HO H H O O H N O

(1) Ac_2O, pyr

(2) mCPBA

(80%)

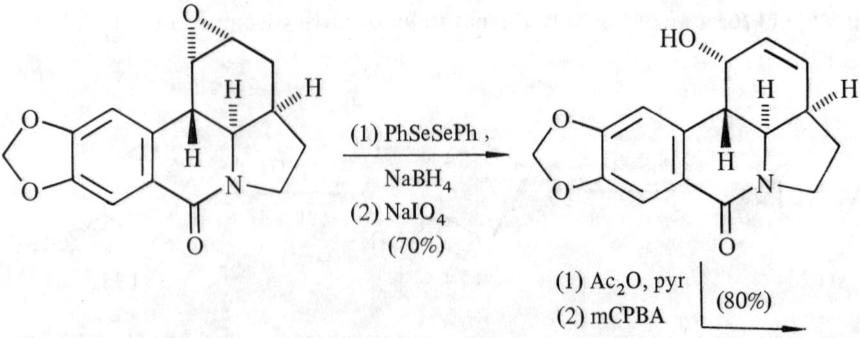

(1) PhSeSePh, $NaBH_4$
(2) $NaIO_4$
(40%)

(1) Ac_2O, pyr
(2) $LiAlH_4$

179

(4.114)

COOEt

$OSiEt_3$

+

Δ
(87%)

EtOOC

$OSiEt_3$

(1) mCPBA
(2) DIBAL
(85%)

$OSiEt_3$

HO

(1) PhSeSePh, $NaBH_4$
(2) mCPBA

$OSiEt_3$

HO

OH

OAc

OAc

BzO

(1) mCPBA
(2) PhCOCl
(3) HCl
(4) Ac_2O

180

(4.115)

COOMe

COOMe

(1) PhSeSePh, $NaBH_4$
(2) H_2O_2
(28%)

COOMe COOH (1) NaOH (2) $H^{\oplus}$ O COOMe OH O COOH OH 181 (4.116)

Still has recently used this reaction sequence in the conversion of (+)-carvone (**182**) to intermediate **183**, which was elaborated to the germacranolide eucannabinolide (**184**, Eq. 4.117).[165]

O 182 (1) $LiAlH_4$ (2) mCPBA (3) $PhCH_2OCH_2Cl$, i-Pr_2NEt (70%) Ph O O O (1) PhSeK, LiBr (2) H_2O_2 (3) Jones oxidation (53%) AcO O O O OH OH O 184 Ph O O O 183 (4.117)

Under the standard conditions for epoxide opening, phenylselenolate anion ($PhSe^{\ominus}$) usually cleaves any methyl esters or acetates present in the substrate (§8.5). By using the ethyl esters, however, and performing the reaction under strictly anhydrous conditions, these side reactions are avoided, as shown in Eq. 4.117a.[233]

EtOOC O AcO (1) PhSeNa, anhydrous conditions (80%) (2) mCPBA (60–65%) OH EtOOC AcO (4.117a)

Trost has examined the conversion of oxaspiropentanes (**185**) to vinyl cyclopropanols (**186**) by means of sodium phenylselenolate. It was found that the β-hydroxyselenides (**187**) could indeed be formed, but oxidation to the selenoxides resulted in carbonium ion formation and rearrangement to the cyclobutanones (**188**, Eq. 4.118). Vinylcyclopropanols (**186**) were obtained by a merged substitution–elimination process, which gave **186** directly from **185** without an oxidation step.[166] This transformation was included in Trost's synthesis of aphidocholin (**20**, Eq. 2.7), outlined in Eq. 4.119.[167]

PhSeSePh, $NaBH_4$, Δ; PhSeSePh, $NaBH_4$; SePh; HO; [O]; O

185 **186** **187** **188**

(4.118)

$\ominus$ $\oplus$ SPh_2; PhSeSePh, $NaBH_4$, Δ; O; H; OH; HO; HO

(4.119)

4.4 *Synthesis of terminal olefins*

The conversion of primary derivatives (**189**) to terminal olefins (**191**) is a very useful synthetic process. Using organoselenium methodology, the conversion can be achieved by transforming **189** to the corresponding aryl selenide (**190**) which affords the terminal olefin (**191**) by oxidation–*syn* elimination (Eq. 4.120). Best results are obtained by using the o-$O_2NC_6H_4Se$ group, which eliminates more readily than the PhSe group.[52]

ArSeM → [O] (4.120)

X SeAr

189 **190** **191**

Aryl selenides (**190**) can be prepared by reacting a bromide (**189**, X = Br), a mesylate (**189**, X = OMs) or a tosylate (**189**, X = OTs) with the nucleophilic ArSeNa, generated *in situ* by sodium borohydride reduction of ArSeCN or ArSeSeAr[52] (Table 4.1, Part B). Synthetic applications of the method include: the synthesis of vernolepin analogs **192** (Eq. 4.121)[168] and **193** (Eq. 4.122),[169] vernolepin (**194**, Eq. 4.123),[170] a clavulanic acid analog (**195**, Eq. 4.124),[171] the alkaloid paliclavine (**196**, Eq. 4.125)[172] and more recently, the antibiotic bicylcomycin.[231]

Br $SeC_6H_4NO_2$-o

o-$O_2NC_6H_4SeCN$

$NaBH_4$

(87%)

O O O O

H OAc COOMe H OAc COOMe

(1) H_2O_2

(2) K_2CO_3

(3) TsOH

(81%)

(4.121)

O O H O O

192

OTs

PhSeSePh, $NaBH_4$

SePh

(1) O_3

(2) Δ

(60%)

193

(4.122)

OMs

PhSeNa

SePh

OEt

OEt

EtOOC

(1) O_3

(2) Δ

(3) CF_3COOH

MeOOC

OH

194

(4.123)

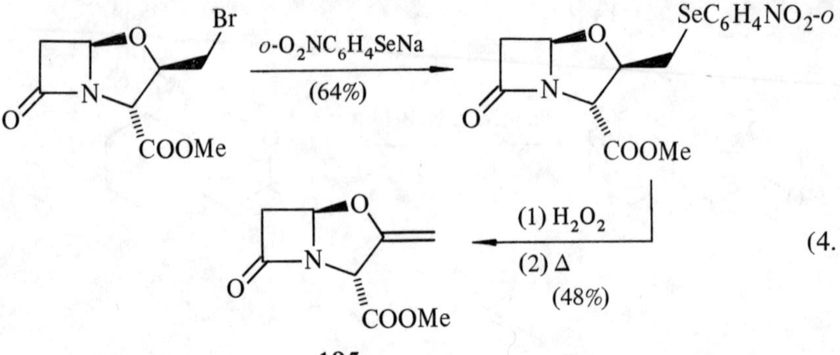

(4.124)

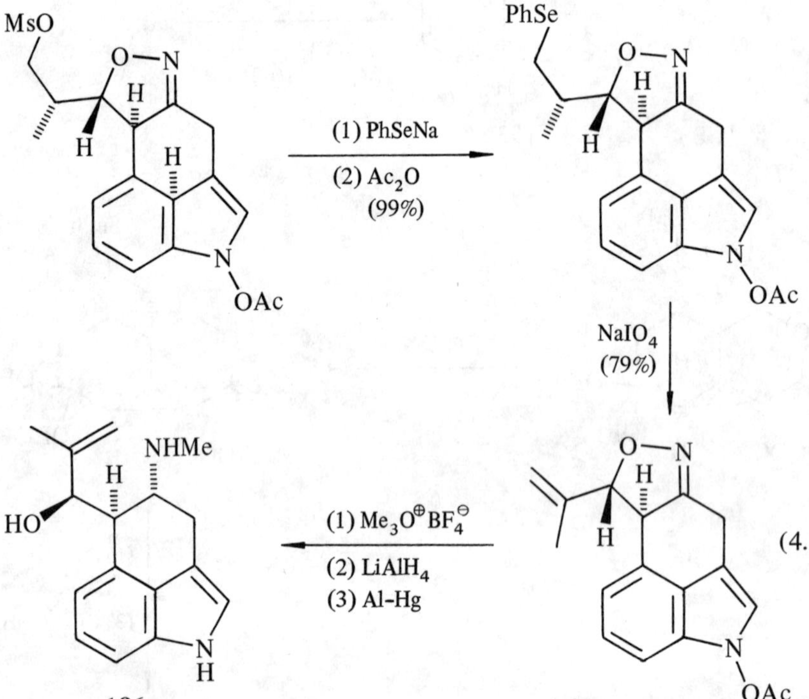

(4.125)

An advantage of this olefin synthesis is that the intermediate selenides (**190**) are quite stable and can survive a variety of non-oxidative and non-basic conditions. Thus, the selenides can be considered to be "protective groups" for the corresponding olefins. The survivability of the PhSe group is illustrated in a recent synthesis of the insect poison pederine (**198**), in which the intermediate selenide (**197**) was carried through ten steps before it was oxidatively removed to form the olefin (Eq. 4.126).[173]

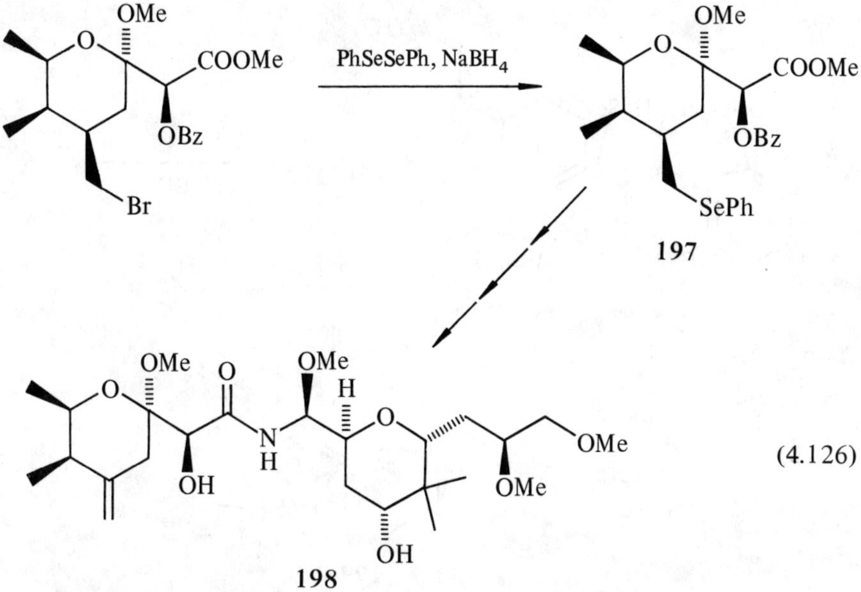

PhSeSePh, $NaBH_4$

197

198

(4.126)

The use of carboxylic esters as precursors to the desired primary selenides allows the synthesis of disubstituted terminal olefins by α-alkylation of the esters. The method was used by Grieco in his elegant synthesis of moenocinol (**199**, Eq. 4.127).[174]

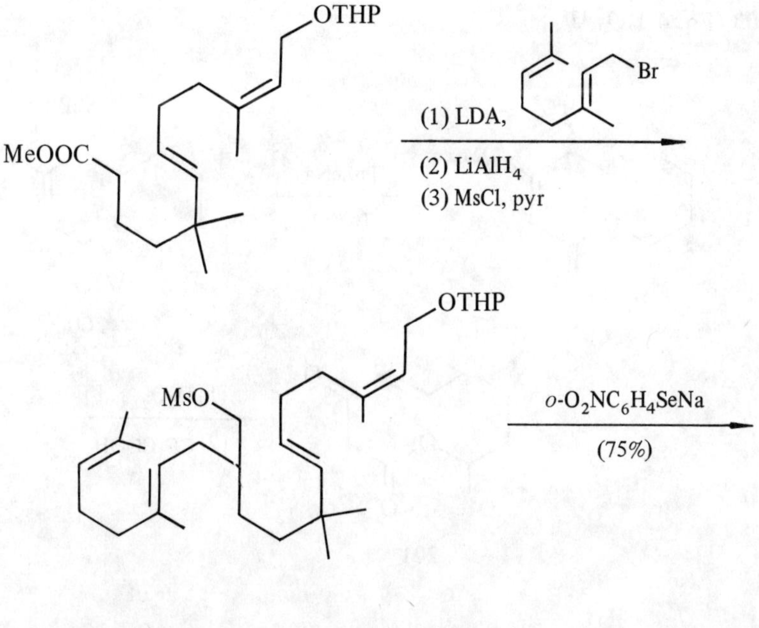

(1) LDA,

(2) $LiAlH_4$

(3) MsCl, pyr

o-$O_2NC_6H_4SeNa$

(75%)

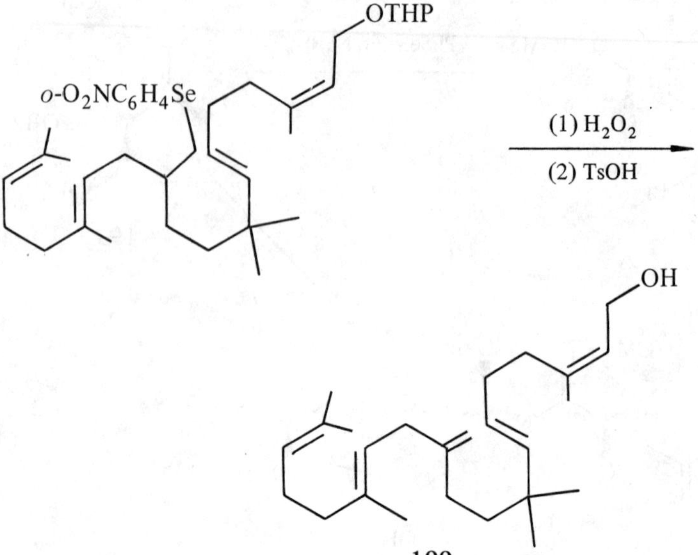

$o\text{-}O_2NC_6H_4Se$ … OTHP → (1) H_2O_2 (2) TsOH → OH

199

(4.127)

Application of the method to the formation of internal olefins is not so common. An example is found in the synthesis of diplodialide A (**201**), shown in Eq. 4.128.[121] Using either diastereomer of selenide **200**, it was possible to get the naturally occurring *trans*-diplodialide A (**201**) or its *cis* isomer. Of course, with smaller rings, only the *cis* olefin is obtained, as in the conversion of **202** to **203** (Eq. 4.129).[175]

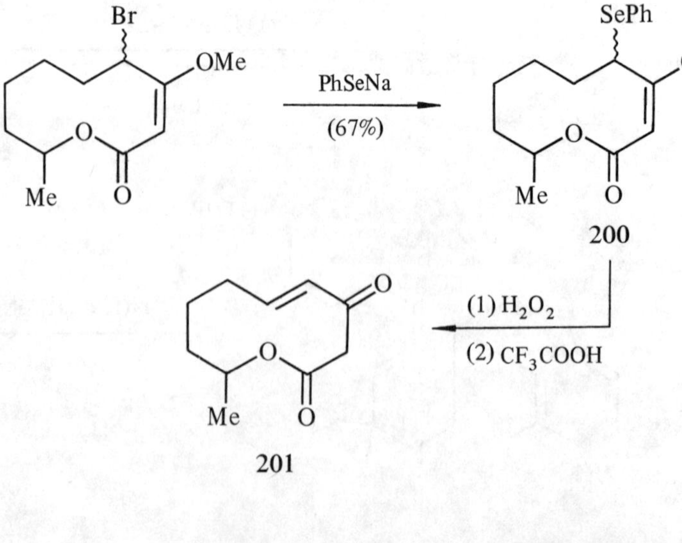

Br, OMe, O, Me, O → PhSeNa (67%) → SePh, OMe, O, Me, O

200

(1) H_2O_2 (2) CF_3COOH

O, O, Me, O

201

(4.128)

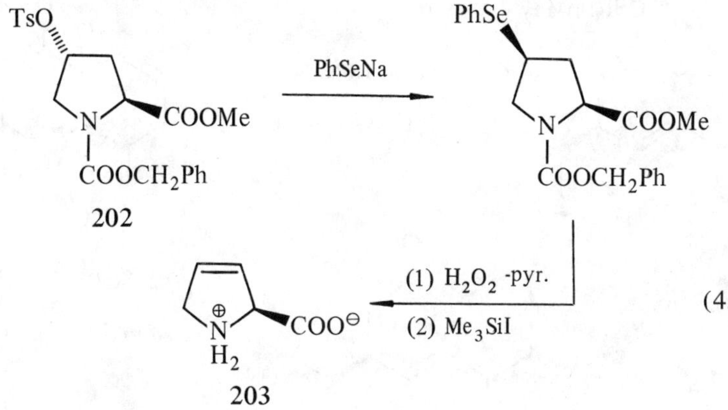

(4.129)

Direct synthesis of selenides of type **190** from primary alcohols (**189**, X = OH) can be accomplished by a method developed by Grieco, which consists in treating the alcohol with *o*-nitrophenylselenocyanate (ArSeCN) in the presence of tributylphosphine (n-Bu_3P). This very efficient method has been used in numerous syntheses, including those of the sex hormone estradiol (**204**, Eq. 4.130),[176] the sex excitant pheromone of the American cockroach, periplanone B (**205**, Eq. 4.131),[177] the alkaloid α-lycorane (**206**, Eq. 4.132),[178] the ionophone antibiotic calcimycin (**207**, Eq. 4.133),[179] the Prelog–Djerassi lactone (**208**, Eq. 4.134),[180] the sesquiterpene sativene (**209**, Eq. 4.135) [181] and the monoterpenes semburin and isosemburin[232] (**224** and **225**, Eq. 4.150).

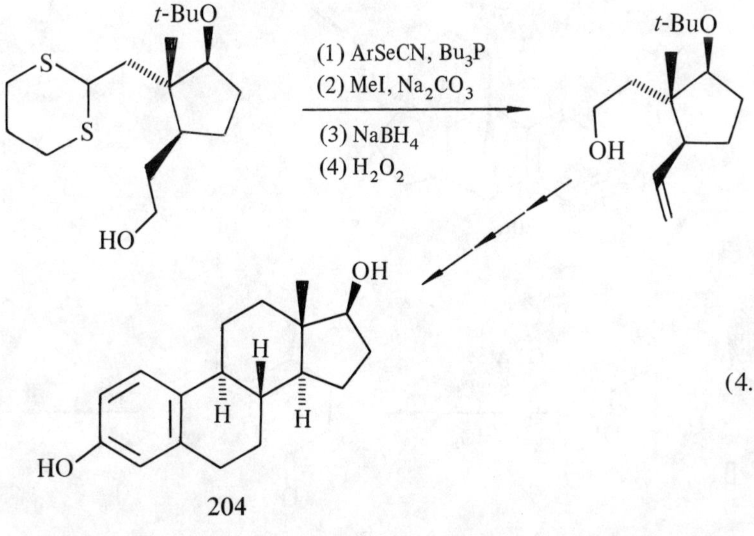

(4.130)

OSi^{t}BuMe$_2$ O HO

(1) ArSeCN, Bu_3P

(2) H_2O_2

(54%)

OSi^{t}BuMe$_2$ O

O O O

205

(4.131)

OH O O N O

(1) ArSeCN, Bu_3P

(2) $NaIO_4$

O O N O

H H O O H N

206

(4.132)

O O OH

(1) ArSeCN,

Bu_3P

(2) $NaIO_4$

O O

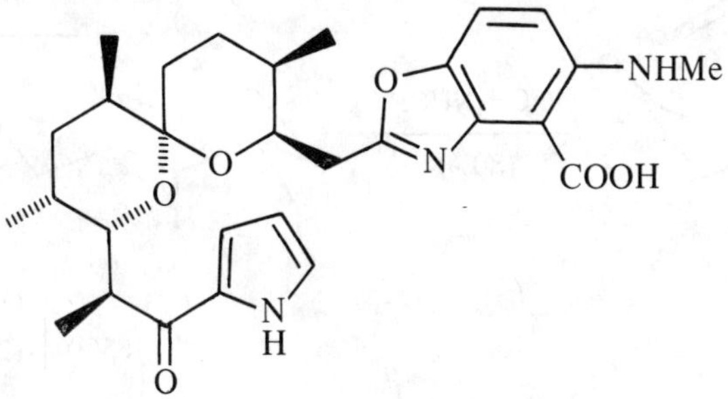

207

(4.133)

(1) ArSeCN, Bu_3P

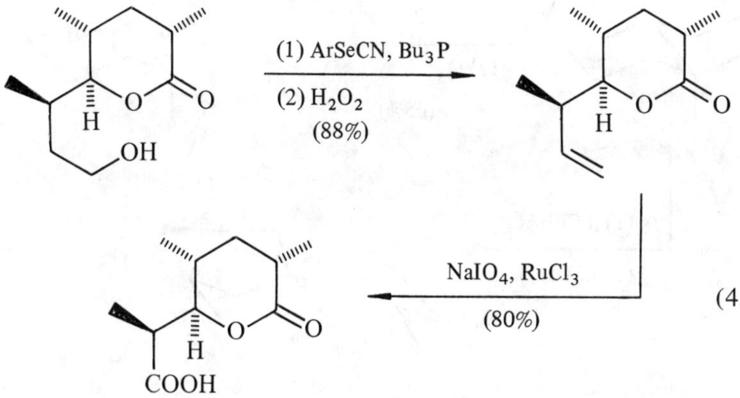

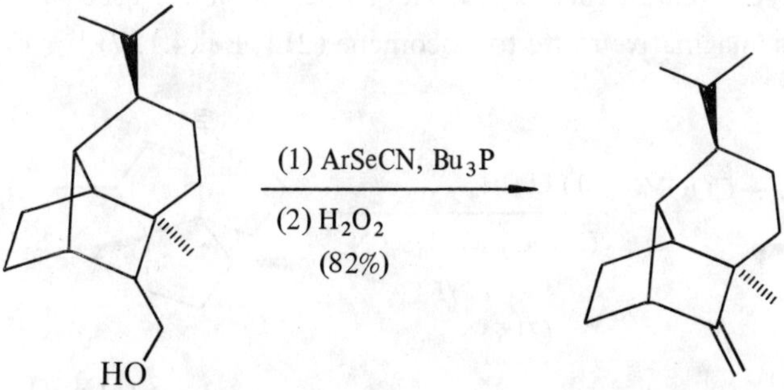

(2) H_2O_2

(88%)

$NaIO_4$, $RuCl_3$

(80%)

208

(4.134)

(1) ArSeCN, Bu_3P

(2) H_2O_2

(82%)

209

(4.135)

The method allowed the application of intramolecular nitrone–olefin cycloadditions to the synthesis of the antifertility agent 7,12-secoishwaran-12-ol (210), outlined in Eq. 4.136.[182]

PhCH$_2$NHOH (80%)

O N Ph

(65%) (1) H_2, Pd$(OH)_2$ (2) NH_2OSO_3H, NaOH, Δ

OH

(1) o-$O_2NC_6H_4SeCN$, Bu_3P (2) O_3 (65%)

(1) mCPBA (2) $LiAlH_4$

OH

(4.136)

210

Application of the method to the isomerization of an exocyclic olefin bond to an endocyclic one further expands its scope. Oppolzer used such an isomerization in his imaginative route to isocomene (**211**, Eq. 4.137).[183]

COOMe (1) $LiAlH_4$ (2) o-$O_2NC_6H_4SeCN$, Bu_3P, pyr (71%)

SeC_6H_4-o-NO_2

(58%) (1) $NaIO_4$ (2) Δ (80° C)

TsOH

(75%)

(4.137)

211

Ozonolysis of the olefin to give the corresponding aldehyde is a further application of the method. An example is found in the early stages of Woodward's synthesis of erythromycin, shown in Eq. 4.138.[184]

(1) Raney Ni

(2) o-$O_2NC_6H_4SeCN$, Bu_3P

(3) H_2O_2

(4) O_3

(80%)

Erythromycin (4.138)

Conversion of a dihydroxy compound to the diselenide (simultaneous conversion of both hydroxyl groups) is not always possible. A stepwise conversion, however, can solve this problem, as in Grieco's synthesis of costunolide (131, Eq. 4.70) from santonin (212), outlined in Eq. 4.139.[137]

212

(1) O_3

(2) $NaBH_4$ (91%)

(1) o-$O_2NC_6H_4SeCN$
Bu_3P, pyr
(2) H_2O_2

(1) o-$O_2NC_6H_4SeCN$, Bu_3P
(2) H_2O_2

(1) LDA– PhSeSePh
(2) H_2O_2
(3) Δ

131
(*Cf.* Eq. 4.70) (4.139)

Similar difficulties were encountered in the related syntheses of temisin (**213**)[185,186] and melitensin (**214**),[186] shown in Eq. 4.140.

R^1 = Me, R^2 = H
R^1 = H, R^2 = Me

(1) ArSeCN, Bu_3P
(2) H_2O_2
(3) ArSeCN, Bu_3P
(4) H_2O_2

(1) ArSeCN, Bu_3P
(2) H_2O_2

LDA

ArSeCN = o-$O_2NC_6H_4SeCN$

(1) SeO_2, t-BuOOH
(2) n-Bu_4NF

R = Si^tBuMe_2
n-Bu_4NF
213, R = H

214

(4.140)

A similar diselenenylation of a diol, however, proceeded quite efficiently in a recent synthesis of β-elemol, as shown in Eq. 4.140a.[226]

HO HO H OH (1) ArSeCN, Bu_3P (84%) (2) $NaIO_4$ (72%) H OH (4.140a)

An important feature of Grieco's method for the preparation of primary selenides of type **190** is that it is sensitive to steric hindrance. Thus, secondary alcohols and hindered primary alcohols react much slower than unhindered primary alcohols. Applications of this selectivity are found in Grieco's synthesis of estradiol (**204**, Eq. 4.130), shown in Eq. 4.141,[187] in the synthesis of a precursor of verrucarol (**102**, Eq. 4.57). shown in Eq. 4.142[188] and in a synthesis of the indole alkaloid antirhine (**215**, Eq. 4.143).[189]

OH MeO OH (1) ArSeCN, Bu_3P (2) H_2O_2 (89%) OH MeO (4.141)

204 (Eq. 4.130) (1) Δ (2) BBr_3

OH OH O (1) ArSeCN, Bu_3P (2) mCPBA (67%) OH O **102** (Eq. 4.57) (4.142)

ArSeCN, Bu_3P (64%)

mCPBA (72%)

(4.143)

215

A more convenient method for the conversion of primary alcohols (**189**, X = OH) to primary selenides (**190**) has been reported by Grieco and Nicolaou.[40] This method replaces ArSeCN by N-phenylselenophthalimide (N–PSP),[39] a readily available, stable, crystalline and odorless reagent. The N–PSP method was used by Ley in his synthesis of clerodane diterpenes,[190] such as ajugarin (**218**) and model compound **216**, shown in Eq. 4.144. Compound **216** is an analog of both ajugarin (**218**) and clerodin (**217**), and was found to have similar insect antifeedant properties as the natural products.[190]

N-PSP, Bu_3P (89%)

(1) O_3
(2) Δ
(3) CF_3COOH
(92%)

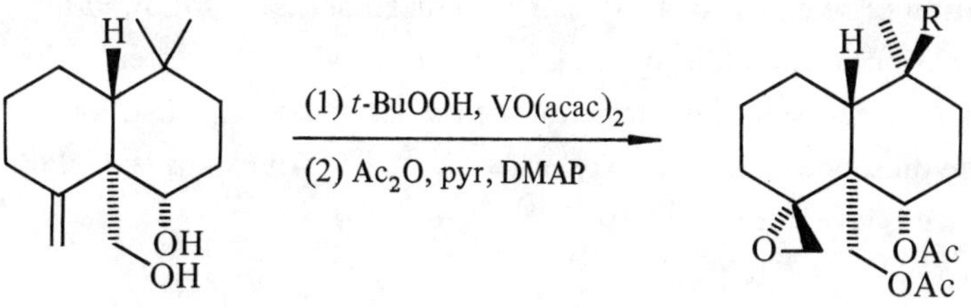

(1) t-BuOOH, VO(acac)$_2$
(2) Ac_2O, pyr, DMAP

(4.144)

216: R = Me

217: R =

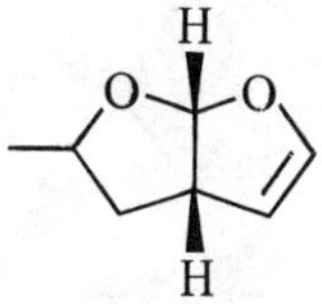

218: R =

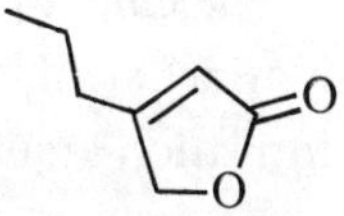

Primary selenides of type **220** can be prepared according to Smith[51, 191] and Liotta[54] by nucleophilic ring opening of lactones (**219**) with phenyl selenide anion (PhSe$^\ominus$, Table 4.1, Part B), as shown in Eq. 4.145. Presumably, the nucleophilic attack of PhSe$^\ominus$ (a *soft* nucleophile) proceeds via an S_N2-type mechanism on the carbinol carbon (*soft–soft* interaction).[54] Thus, the reaction is sensitive to steric hindrance around the carbinol center and it does not proceed very well when this center is secondary or tertiary.[51, 54, 191]

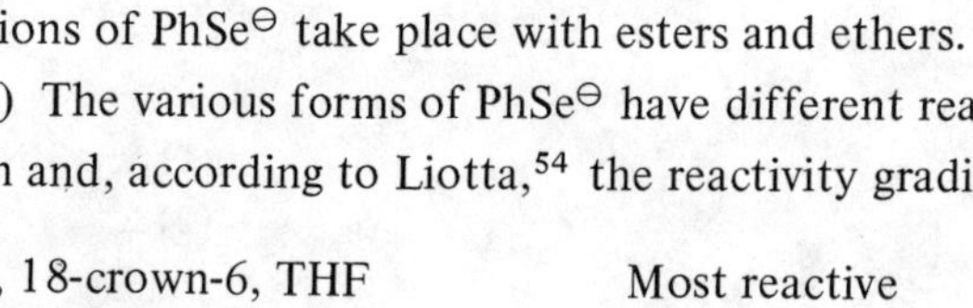

PhSe$^\ominus$

(4.145)

219 **220**

Similar reactions of PhSe$^\ominus$ take place with esters and ethers. (They are discussed in §8.3.) The various forms of PhSe$^\ominus$ have different reactivity in this type of reaction and, according to Liotta,[54] the reactivity gradient is as follows:

PhSeH, NaH, 18-crown-6, THF	Most reactive
PhSeH, NaH, HMPA, THF	↑
PhSeLi, HMPA, THF	
PhSeLi, THF or Et_2O	
PhSeSePh, $NaBH_4$, THF or EtOH	Least reactive

Esterification of selenides of type **220** with diazomethane, followed by oxidative elimination of the PhSe group, gives the ω-vinyl methyl esters. The esterification step is necessary here because the oxidation-*syn* elimination will not occur in the presence of the carboxyl group.[51, 54, 191] An example of the reaction sequence is given in Eq. 4.146.[51, 191]

PhSeSePh, $NaBH_4$ (90%); (1) CH_2N_2 (2) O_3 (3) Pyr, Δ (71%) — OH, SePh, OMe (4.146)

The mildness of the method is indicated in the following example (Eq. 4.147), in which neither the ketone nor the methyl ether is affected. During the transformation sequence, however, an epimerization of the ring-junction proton takes place, probably during the selenoxide elimination step.

H, O, OMe; (1) PhSeNa, HMPA (2) CH_2N_2 (3) O_3 (4) Et_2NH, Δ (73%) — H, O, MeO, OMe (4.147)

The method is also useful in the synthesis of diene esters, such as **221** (Eq. 4.148).[192]

OH, O, H; (1) PhSeSePh, K (2) CH_2N_2 (62%) — OH, SePh, COOMe, H; (1) AcOOH (2) Δ (73%) — OH, COOMe, H (4.148)

221

In the case of β-keto lactones, it is possible to eliminate the resulting carboxyl group, as in the synthesis of dehydrojasmone (222, Eq. 4.149).[193] Alternatively, the carboxyl group can be transformed into other functional groups prior to

$EtC{\equiv}CCH_2Br$, K_2CO_3 (90%)

(1) PhSeH, NaH, HMPA
(2) Δ
(97%)

SePh

mCPBA
(85%)

222

(4.149)

elimination of the PhSe group, as in the synthesis of antirhine (215, Eq. 4.143) shown in Eq. 4.150. [194] It is interesting to note that ozonolysis of the selenide (223) resulted in simultaneous double bond fission and double bond formation! The intermediate selenide lactone (223) was also used by the same group of researchers in a synthesis of semburin (224) and isosemburin (225), also shown in Eq. 4.150.[195]

CN H PhSeSePh, Na

CN COOH H SePh

(1) ClCOOEt, Et_3N, $NaBH_4$
(2) KOH
(79%)

223

O_3
(61%)

CHO

215 (Eq. 4.143)

(1) $NaBH_3CN$ tryptamine
(2) DIBAL
(3) HCl

(1) DIBAL
(2) $NaBH_4$

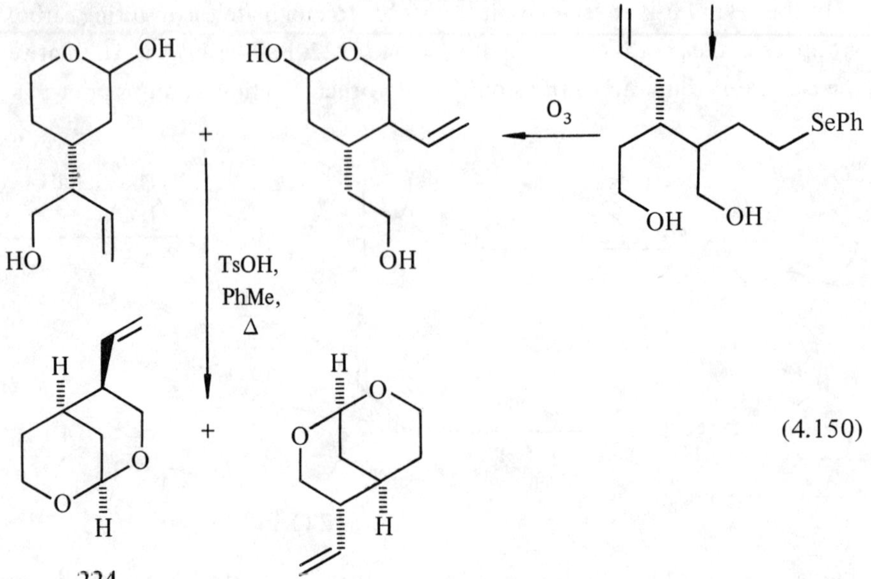

(4.150)

224

225

Conversion of the carboxyl group to a methyl carbamate group through the Curtius rearrangement, is found in a recent synthesis of the carbocyclic nucleoside antibiotic neplanocin A (**226**), shown in Eq. 4.151.[196]

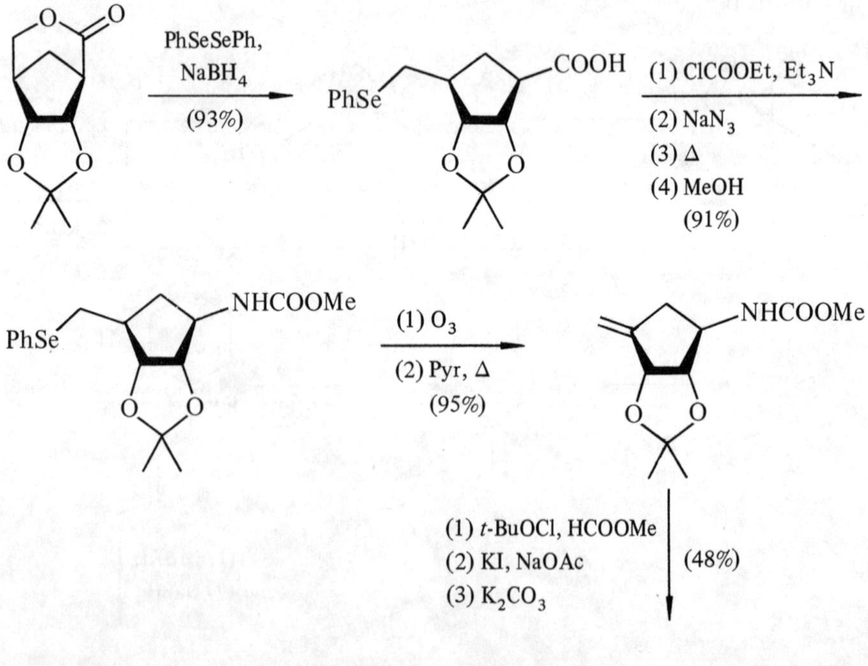

(4.151)

226

α-Aryloxy-substituted butyrolactones, upon PhSeNa-cleavage, esterification and oxidation–*syn* elimination, are transformed into substrates that undergo the Claisen rearrangement. The method was included in an efficient, practical and convenient synthesis of the antibiotic nanaomycin (**227**), summarized in Eq. 4.152.[197]

(1) $Na_2S_2O_4$
(2) $Me_2C(OMe)_2$, BF_3
(3) Br-butyrolactone, CsF, Δ
(74%)

(4.152)

(81%) (1) PhSeNa (2) CH_2N_2

(1) H_2O_2 (2) Δ (72%)

(91%) Na_2CO_3

(1) Ag_2O
(2) Zn–HCl, MeCHO, Δ
(3) Ag_2O
(4) H_2SO_4
(5) KOH

HO O O COOMe O COOMe O O O

227

Formation of the primary selenide **229** by Michael addition of $PhSe^{\ominus}$ to α-methylene lactones **228**, was considered as a method for the protection of this sensitive functionality, since the selenide (**229**) reforms the α-methylene lactone (**228**) by oxidative elimination (Eq. 4.153).[50]

"PhSeH" [O] SePh

228 229 (4.153)

An application of the method is found in a synthesis of the pheromone ipsenol (**231**, Eq. 4.154).[198] The PhSe group of **230**, however, was eliminated non-oxidatively, probably through a retro-Michael reaction.

PhSeSePh, $NaBH_4$ (100%) SePh

(100%) DIBAL

OH $Ph_3P{=}CH_2$ (25%) OH SePh

231 230 (4.154)

Another use for the nucleophilic phenylseleno species ($PhSe^{\ominus}$) is in the ring opening of activated cyclopropanes, *e.g.* cleavage of **232** to form primary selenides (**233**, Eq. 4.155), reported by Smith.[56]

PhSe$^{\ominus}$ (4.155)

SePh

232 233

Lithium phenylselenolate (PhSeLi) was found to be more effective for the cleavage of cyclopropyl ketones (Eq. 4.156), while the sodium phenylselenolate worked better with nitriles (Eq. 4.157).

PhSeH, *n*-BuLi (64%) SePh (4.156)

CN PhSeSePh, $NaBH_4$ (47%) NC SePh (4.157)

Masamune employed the method to prepare several useful synthetic intermediates, such as **234** (Eq. 4.158).[199]

OH HOOC (1) Cyclopropyl lithium (2) $Me_2{}^tBuSiCl$ (83%) OSi^tBuMe_2 O

PhSeLi, 18-C-6, Δ (91%)

(1) $(c\text{-}C_5H_9)_2BOTf$, $i\text{-}Pr_2NEt$
(2) EtCHO
(97%)

(86%) (1) HF (2) O_3

(1) $NaIO_4$
(2) CH_2N_2
(3) $Me_2{}^tBuSiOTf$

234

(4.158)

Incorporation of a primary selenide functionality as a precursor to the *N*-vinyl group has been achieved by using the novel selenium-containing reagent 2-phenylselenylethylamine ($PhSeCH_2CH_2NH_2$).[200] The method was used by Heck in an approach to β-lactam antibiotics, as shown in Eq. 4.159.[200] This selenium reagent was also used recently by Magnus, in his synthetic studies on indole alkaloids.[227]

(1) $PhSeCH_2CH_2NH_2 \cdot MgSO_4$
(2) N_3CH_2COCl, Et_3N
(93%)

(1) mCPBA
(2) $i\text{-}Pr_2NH$
(99%)

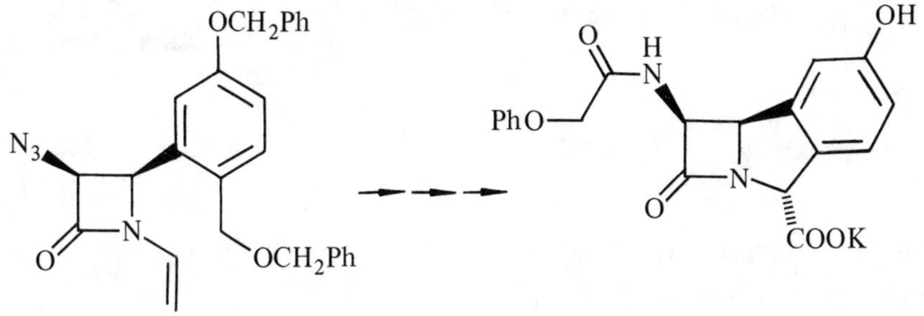

(4.159)

4.5 [2,3]-Sigmatropic rearrangements of β,γ-unsaturated selenoxides

Allylic selenoxides (**235**) rearrange readily to form allylic alcohols (**236**, Eq. 4.160). This facile [2,3]-sigmatropic rearrangement was first observed by Sharpless during mechanistic studies on allylic oxidation with selenium dioxide[201] and during the conversion of allylic epoxide **237** to *trans*-allylic diol **238** (Eq. 4.161).[3]

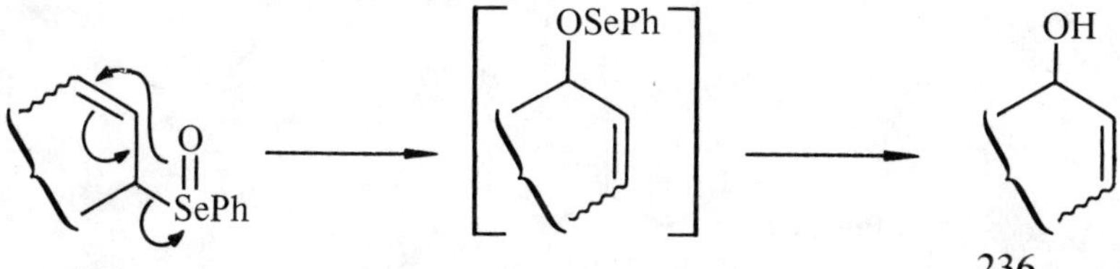

235 **236**

(4.160)

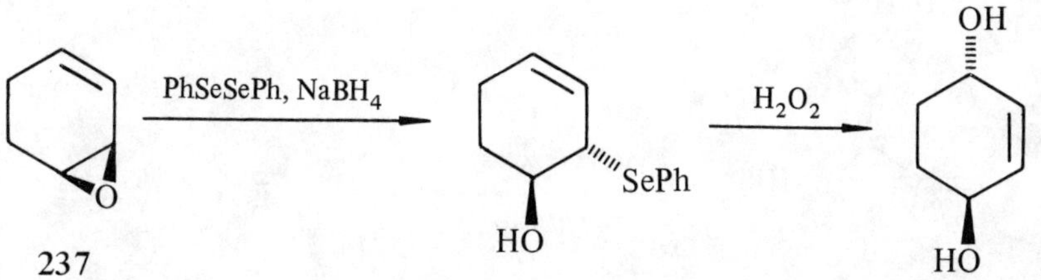

237 **238**

(4.161)

Reich has exploited this process for the preparation of allylic alcohols (**236**),[202] while Clive has used it for the 1,3-transposition of primary allylic alcohols, *e.g.* **239**→**240** (Eq. 4.162).[203]

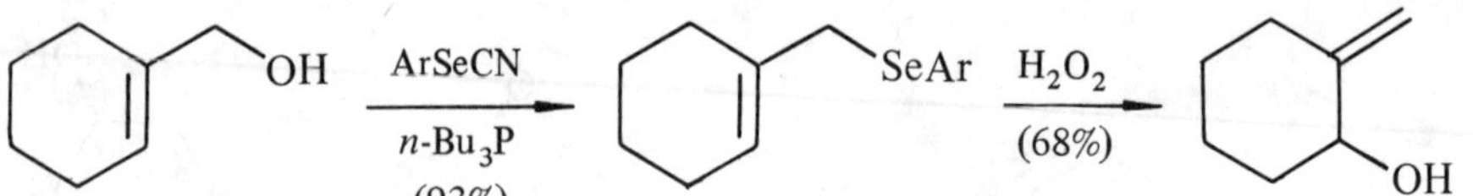

OH
ArSeCN
n-Bu3P
(93%)
SeAr
H2O2
(68%)
OH

239 240

(4.162)

The rearrangement is particularly favored in systems where the *syn*-elimination is restricted, as in the example shown in Eq. 4.163.[204]

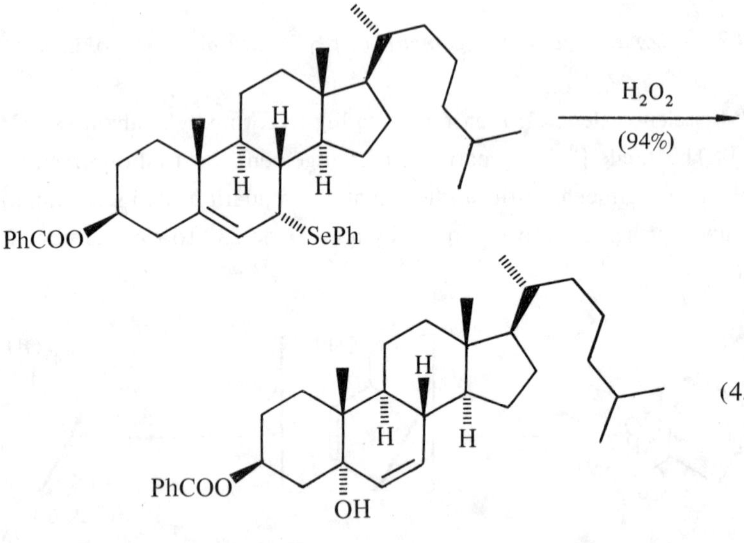

H
H
H
PhCOO
SePh
H2O2
(94%)
H
H
H
PhCOO
OH

(4.163)

An application of this [2,3]-shift is found in the synthesis of litsenolide C (**241**, Eq. 4.164).[205]

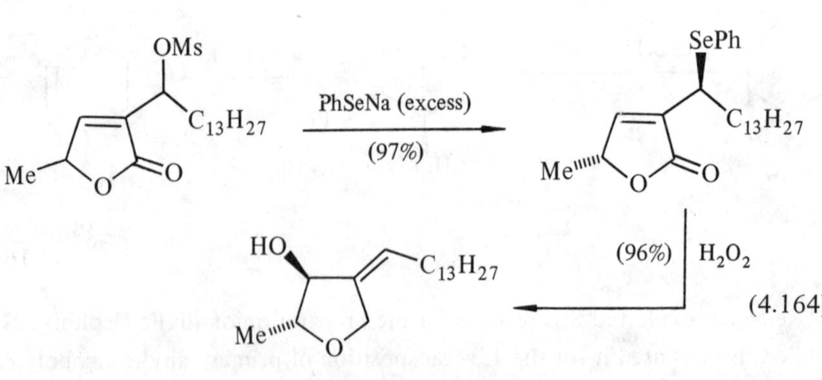

OMs
C13H27
Me
O
O
PhSeNa (excess)
(97%)
SePh
C13H27
Me
O
O
(96%)
H2O2
HO
C13H27
Me
O

(4.164)

241

Similar [2,3]-sigmatropic rearrangements occur with allenyl selenoxides (**242**) to form propargyl alcohols (**243**, Eq. 4.165),[207] and with propargyl selenoxides (**244**) to form enones (**245**, Eq. 4.166).[207]

OSePh OH (4.165)

Ph

242 **243**

OSePh O X (4.166)

Ph

X = H, SePh

244 **245**

The reverse rearrangement, *i.e.* conversion of an allylic alcohol to an allylic selenoxide, followed by selenoxide *syn*-elimination, leads to dienes. This process, devised by Reich, is exemplified in Eq. 4.167.[208] However, better results were obtained with the corresponding sulfur reagents.

R OH ArSeCl (65%) R OSeAr R SeAr O

R

(4.167)

Allylic selenides can also be prepared from allylsilanes, thereby opening some more routes to allylic alcohols, as shown in Eq. 4.168[209] and Eq. 4.169.[210] These reactions proceed via addition of PhSeCl to the olefin, followed by dechlorosilylation and isomerization to the most stable selenide.[209]

Me$_3$Si–CH$_2$CH=CHCH$_2$CH$_2$Ph → (1) PhSeCl, (2) $SnCl_2$ cat. (97%) → PhSe–CH$_2$CH=CHCH$_2$CH$_2$Ph → H_2O_2 (91%) → CH$_2$=CHCH(OH)CH$_2$CH$_2$Ph (4.168)

Br–CH$_2$CH=CHCH$_2$SiMe$_3$ → (1) Mg–CuI, (2) Propylene oxide, (3) Ac_2O–pyr (83%) → CH$_3$CH(OAc)CH$_2$CH=CHCH$_2$SiMe$_3$ → (1) PhSeCl, (2) $SnCl_2$ cat (91%) → CH$_3$CH(OAc)CH$_2$CH=CHCH$_2$SePh → H_2O_2 (85%) → CH$_3$CH(OAc)CH$_2$CH(OH)CH=CH$_2$ (4.169)

An application of the reaction to the synthesis of iridoids was reported recently by Hutchinson (Eq. 4.170).[211]

5-(Trimethylsilyl)cyclopentadiene (SiMe$_3$) + HO–CH=C(COOMe)–CHO → (1) MeCN, −10° C, (2) $BF_3 \cdot Et_2O$, MeOH, 0° C → bicyclic iridoid (COOMe, Me$_3$Si, OMe) → PhSeCl, 0° C (50–60%) →

PhSe COOMe → (H_2O_2, Pyr, 0° C) → COOMe, HO, OMe (42%) + COOMe, O, OMe (30%) (4.170)

Allyl silanes are converted directly to allylic selenoxides, which subsequently rearrange to allylic alcohols, by treatment with benzeneseleninic anhydride [PhSe(O)OSe(O)Ph], as indicated in Eq. 4.171.[212]

PhSe(O)–O–Se(O)Ph, $SiMe_3$, OAc → [PhSe=O, OAc] → OH, OAc (4.171)

The [2,3]-sigmatropic rearrangement is not a favored process when the olefin of the allylic selenide is conjugated with a carbonyl group. In that case, the *syn*-elimination to give the conjugated diene, is the more favored pathway. Typical examples are given in Eq. 4.49,[116] 4.128,[121] 4.172[213] and 4.173.[214]

NC CN Ph Ph → (LDA–PhSeBr, (66%)) → NC CN PhSe Ph Ph → (H_2O_2, pyr, (57%)) → NC CN Ph Ph (4.172)

LDA / PhSeBr ; H_2O_2 ; SePh

(4.173)

4.6 *Eliminations of β-hydroxyselenides to form olefins*

The development of the various ways for the synthesis of β-hydroxyselenides (**246**) from olefins (§4.2), epoxides (§4.3) and selenium-containing synthons (§6), has greatly increased the usefulness of the hydroxyselenides as synthetic intermediates. Their conversion to allylic alcohols has already been discussed (§4.2 and §4.3). Another important transformation of β-hydroxyselenides is their conversion to olefins(**247**)by elimination of the elements of "PhSeOH" (Eq. 4.174). This is achieved by transforming the hydroxyl group to a better leaving group with a variety of reagents (Table 4.5).

SeR ; OH ; **246** ; **247**

(4.174)

Table 4.5. Methods for the conversion of β-hydroxyselenides to olefins.

Reagent	*Ref.*
$MeSO_2Cl–Et_3N$	215
TsOH or $HClO_4$	216
$(CF_3CO)_2O–Et_3N$	216
$SOCl_2–Et_3N$	216
$PI_3–Et_3N$	217
$Imidaz_2CO$	217
$Me_2{}^tBuSiCl$ and $Na–NH_3$ (*l*)	218
$Me_3SiCl–NaI$	219

Typical examples are shown in Eq. 4.175,[215] 4.176,[217] 4.177[218] and 4.178.[219]

SePh, OH; Ph — $MsCl{-}Et_3N$ (73%) → Ph (4.175)

SeMe, Me, C_9H_{19}, OH — PI_3, Et_3N (58%) → Me, C_9H_{19} (4.176)

OSi^tBuMe_2, SePh — $Na{-}NH_3$ (l) (75%) → (4.177)

SePh, OH — Me_3SiCl, NaI (90%) → (4.178)

Krief has combined the reaction with the opening of epoxides with $RSe^{\ominus}$ (Eq. 4.111) as a means of stereospecific *cis*-deoxygenation of epoxides, *e.g.*, **248** → **249** → **250** (Eq. 4.179).

H, O, H, R, R — $MeSe^{\ominus}$ → H, OH, H, R, SeMe, R — $SOCl_2{-}NEt_2$ → H, H, R, R (4.179)

248 **249** **250**

Also, the reaction was combined with several other routes to hydroxyselenides, such as addition of Grignard reagents to α-phenylseleno aldehydes (**251**, §4.1.2), shown in Eq. 4.180,[220] or reduction of α-phenylseleno ketones (**253**. §4.1.1), shown in Eq. 4.181.[220] The first sequence (Eq. 4.180) leads primarily to *E*-olefins (**252**), while the second (Eq. 4.181) gives *Z*-olefins (**254**) as the major products.[220]

R′MgBr; TsOH

251 **252**

(4.180)

$LiAlH_4$; TsOH

253 **254**

(4.181)

An organoselenium-mediated, highly stereoselective elimination of *vic*-dihalides or β-halogenoselenides has also been reported by Krief.[221] Examples are shown in Eq. 4.182 and Eq. 4.183.

PhSeNa (96%)

$R = C_9H_{17}$

(100% trans)

(4.182)

PhSeNa (95%)

(100% cis)

(4.183)

REFERENCES

1. Jones, D. N., Mundy, D., and Whitehouse, R. D., *J. C. S. Chem. Comm.*, 86 (1970).
2. Walter, R., and Roy, J., *J. Org. Chem.*, **36**, 2561 (1971).
3. Sharpless, K. B., and Lauer, R. F., *J. Am. Chem. Soc.,* **95**, 2697 (1973).
4. Sharpless, K. B., Young, M. W., and Lauer, R. F., *Tetrahedron Lett.*, 1979 (1973).
5. Sharpless, K. B., Lauer, R. F., and Teranishi, A. Y., *J. Am. Chem. Soc.,* **95**, 6137 (1973).
6. Sharpless, K. B., and Lauer, R. F., *J. Org. Chem.,* **39**, 429 (1974).
7. Sharpless, K. B., Gordon, K. M., Lauer, R. F., Patrick, D. W., Singer, S. P., and Young, M. W., *Chem. Scr.,* **8A**, 9 (1975).
8. Reich, H. J., Reich, I. L., and Renga, J. M., *J. Am. Chem. Soc.,* **95**, 5813 (1973).
9. Reich, H. J., Renga, J. M., and Reich, I. L., *J. Org. Chem.,* **39**, 2133 (1974).
10. Reich, H. J., *J. Org. Chem.,* **39**, 428 (1974).
11. Reich, H. J., Renga, J. M., and Reich, I. L., *J. Am. Chem. Soc.,* **97**, 5434 (1975).
12. Reich, H. J., "Organoselenium Oxidations," in Trahanovsky, W. S. (ed.), *Oxidation in Organic Chemistry, Part C*, Academic Press, New York, 1978, Chap. 1, p. 1.
13. Reich, H. J., *Acc. Chem. Res.,* **12**, 22 (1979).
14. Clive, D. L. J., *Tetrahedron,* **34**, 1049 (1978).
15. Clive, D. L. J., *Aldrichimica Acta,* **11**, 43 (1978).
16. Krief, A., *Tetrahedron,* **36**, 2531 (1980).
17. Nicolaou, K. C., *Tetrahedron,* **37**, 4079 (1981).
18. Liotta, D., and Zima, G., *Tetrahedron Lett.,* 4977 (1978).
19. (a) Liotta, D., Zima, G., and Saindane, J., *J. Org. Chem.,* **47**, 1258 (1982); (b) Liotta, D., and Zima, G., *J. Org. Chem.,* **45**, 2551 (1980).
20. Filer, C. N., Ahern, D., Fazio, R., and Shelton, E. J., *J. Org. Chem.,* **45**, 1313 (1980).
21. Raucher, S., *Tetrahedron Lett.,* 3909 (1977).
22. Raucher, S., *J. Org. Chem.,* **42**, 2950 (1977).
23. Denis, J. N., Vicens, J., and Krief, A., *Tetrahedron Lett.,* 2697 (1979).
24. Toshimitsu, A., Uemura, S., and Okano, J., *J. C. S. Chem. Comm.,* 87 (1982).

25. Clive, D. L. J., *J. C. S. Chem. Comm.,* 695 (1973).
26. Clive, D. L. J., *J. C. S. Chem. Comm.,* 100 (1974).
27. Kozikowski, A. P., Sorgi, K. L., and Schmiering, R. J., *J. C. S. Chem. Comm.,* 477 (1980).
28. (a) Toshimitsu, A., Aoai, T., Uema, S., and Okano, M., *J. Org. Chem.,* **45**, 1953 (1980); (b) Toshimitsu, A., Uema, S., and Okano, M., *J. C. S. Chem. Comm.,* 166 (1977).
29. Grieco, P. A., Gilman, S., and Nishizawa, M., *J. Org. Chem.,* **41**, 1485 (1976).
30. Grieco, P. A., and Yokoyama, Y., *J. Am. Chem. Soc.,* **99**, 5210 (1977).
31. Grieco, P. A., Yokoyama, Y., and Williams, E., *J. Org. Chem.,* **43**, 1283 (1978).
32. Tomoda, S., Takeuchi, Y., and Nomura, Y., *J. C. S. Chem. Comm.,* 871 (1982); *Tetrahedron Lett.,* **23**, 1361 (1982); *Chem. Lett.,* 1733 (1982).
33. Hori, T., and Sharpless, K. B., *J. Org. Chem.,* **43**, 1689 (1978).
34. Reich, H. J., Wollowitz, A., Trend, J. E., Chow, F., and Wendelborn, D. F., *J. Org. Chem.,* **43**, 1697 (1978).
35. Labar, D., Krief, A., and Hevesi, L., *Tetrahedron Lett.,* 3697 (1978).
36. Toshimitsu, A., Aoai, T., Owada, H., Uemura, S., and Okano, M., *J. C. S. Chem. Comm.,* 412 (1980).
37. Shimizu, M., Takeda, R., and Kuwajima, I., *Bull. Chem. Soc., Japan,* **54**, 3510 (1981); *Tetrahedron Lett.,* 419, 3461 (1979).
38. (a) Hori, T., and Sharpless, K. B., *J. Org. Chem.,* **44**, 4208 (1979); (b) Frejd, T., and Sharpless, K. B., *Tetrahedron Lett.,* 2239 (1978).
39. Nicolaou, K. C., Claremon, D. A., Barnette, W. E., and Seitz, S. P., *J. Am. Chem. Soc.,* **101**, 3704 (1979).
40. Grieco, P. A., Jaw, J. Y., Claremon, D. A., and Nicolaou, K. C., *J. Org. Chem.* **46**, 1215 (1981).
41. Scarborough, R. M., Jr., Smith, A. B., III, Barnette, W. E., and Nicolaou, K. C., *J. Org. Chem.,* **44**, 1742 (1979).
42. Reich, H. J., and Renga, J. M., *J. Org. Chem.,* **40**, 3313 (1975).
43. Reich, H. J., Renga, J. M., and Trend, J. E., *Tetrahedron Lett.,* 2217 (1976).
44. (a) Toshimitsu, A., Aoai, T., Owada, H., Uemura, S., and Okano, M., *J. Org. Chem.,* **46**, 4727 (1981); (b) Toshimitsu, A., Aoai, T., Uemura, S., and Okano M., *J. C. S. Chem. Comm.,* 1041 (1980); (c) Toshimitsu, A., Owada, H., Aoai, T., Uemura, S., and Okano, M., *J. C. S. Chem. Comm.,* 546 (1981).
45. Hayama, T., Tomoda, S., Takeuchi, Y., and Nomura, Y., *Tetrahedron Lett.,* 23, 4733 (1982).

46. Back, T. G., and Collins, S., *Tetrahedron Lett.,* **21**, 2215 (1980).
47. Gancarz, R. A., and Kice, J. L. *Tetrahedron Lett.,* **21**, 4155 (1980).
48. Miura, T., and Kobayashi, M., *J. C. S. Chem. Comm.,* 438 (1982).
49. Kang, Y.-H., and Kice, J. L., *Tetrahedron Lett.,* **23**, 5373 (1982).
50. Grieco, P. A., and Miyashita, M., *Tetrahedron Lett.,* 1869 (1974).
51. Scarborough, R. M., Jr., and Smith, A. B., III, *Tetrahedron Lett.,* 4361 (1977).
52. Sharpless, K. B., and Young, M. W., *J. Org. Chem.,* **40**, 947 (1975).
53. Liotta, D., Markiewicz, W., and Santiesteban, H., *Tetrahedron Lett.,* 4365
54. (a) Liotta, D., and Santiesteban, H., *Tetrahedron Lett.,* 4369 (1977); (b) Liotta, D., Sunay, U., Santiesteban, H., and Markiewicz, W., *J. Org. Chem.,* **46**, 2605 (1981).
55. Comaseto, J. V., Ferreira, J. T. B., Brandt, C. A., and Petragnani, N., *J. Chem. Res. (S),* 212 (1982).
56. Smith, A. B., III, and Scarborough, R. M., Jr., *Tetrahedron Lett.,* 1649 (1978).
57. Detty, M. R., and Wood, G. P., *J. Org. Chem.,* **45**, 80 (1980).
58. Clive, D. L. J., and Menchen, S. M., *J. Org. Chem.,* **44**, 1883, 4279 (1979); *J. C. S. Chem. Comm.,* 356 (1978).
59. Detty, M. R., *Tetrahedron Lett.,* 5087 (1978).
60. Detty, M. R., *Tetrahedron Lett.,* 4189 (1979).
61. Detty, M. R., and Seidler, M. D., *J. Org. Chem.,* **46**, 1283 (1981).
62. Liotta, D., Paty, P. B., Johnston, J., and Zima, G., *Tetrahedron Lett.,* 5091 (1978).
63. Miyoshi, N., Ishii, H., Kondo, K., Murai, S., and Sonoda, N., *Synthesis,* 300 (1979).
64. Suzuki, M., Kawagishi, T., and Noyori, R., *Tetrahedron Lett.,* 22, 1809 (1981).
65. Dumont, W., and Krief, A., *Angew. Chem. Int. Ed. Engl.,* **14**, 350 (1975).
66. Tieco, M., Testaferri, L., Tingoli, M., Chianelli, D., and Montanucci, M., *Synth. Comm.,* **13**, 617 (1983).
67. Toshimitsu, A., Owada, H., Uemura, S., and Okano, M., *Tetrahedron Lett.,* 23, 2105 (1982); *Tetrahedron Lett.,* **21**, 5037 (1980).
68. Ayrey, G., Barnard, D., and Woodbridge, D. T., *J. Chem. Soc.,* 2089 (1962).
69. Reich, H. J., and Shah, S. K., *J. Am. Chem. Soc.,* **97**, 3250 (1975).
70. Masuyama, Y., Ueno, Y., and Okawara, M., *Chem. Lett.,* 835 (1977).
71. Detty, M. R., *J. Org. Chem.,* **45**, 274 (1980).

72. Vedejs, E., Mullins, M. J., Renga, J. M., and Singer, S. P., *Tetrahedron Lett.*, 514 (1978).
73. Davis, F. A., Stringer, O. D., and Billmers, J. M., *Tetrahedron Lett.*, **24**, 1213 1983).
74. Grieco, P. A., Majectich, G. F., and Ohfune, Y., *J. Am. Chem. Soc.*, **104**, 4226 (1982); *ibid.*, **99**, 7393 (1977).
75. (a) Lansbury, P. T., Hangauer, D. G., Jr., and Vacca, J. P., *J. Am. Chem. Soc.*, **102**, 3964 (1980); (b) for an example of selenenylation via the enolate, see: Ziegler, F. E., and Fang, J. M., *J. Org. Chem.*, **46**, 825 (1981).
76. Danishefsky, S., Vaughan, K., Gadwood, R., and Tsuzuki, K., *J. Am. Chem. Soc.*, **103**, 4136 (1981); *ibid.*, **102**, 4262 (1980).
77. Korzeniowski, S. H., Vanderbilt, D. P., and Hendry, L. B., *Org. Prep. Proc. Int.*, **8**, 81 (1976).
78. Stork, G., and Raucher, S., *J. Am. Chem. Soc.*, **98**, 1583 (1976).
79. Paquette, L. A., and Han, Y. K., *J. Am. Chem. Soc.*, **103**, 1835 (1981); *J. Org. Chem.*, **44**, 4014 (1979).
80. Caine, D., and Frobese, A. S., *Tetrahedron Lett.*, 3107 (1977).
81. Shibasaki, M., Iseki, K., and Ikegami, S., *Tetrahedron Lett.*, **21**, 3587 (1980).
82. Mehta, G., Reddy, A. V., Murthy, A. N., and Reddy, D. S., *J. C. S. Chem. Comm.*, 540 (1982).
83. Annis, G. D., and Paquette, L. A., *J. Am. Chem. Soc.*, **104**, 4504 (1982).
84. (a) Semmelhack, M. F., Tomoda, S., Nagaoka, H., Boettger, S., and Hurst, K. M., *J. Am. Chem. Soc.*, **104**, 747 (1982); (b) Semmelhack, M. F., Tomoda, S., and Hurst, K. M., *J. Am. Chem. Soc.*, **102**, 7567 (1980).
85. Zipkin, R. E., Natale, N. R., Taffer, I. M., and Hutchins, R. O., *Synthesis*, 1035 (1980).
86. Oppolzer, W., and Battig, K., *Helv. Chim. Acta*, **64**, 2489 (1981).
87. Hayashi, Y., Matsumoto, T., Nishizawa, M., Tagami, M., Hyono, T., Nishikawa, N., Uemura, M., and Sakan, T., *J. Org. Chem.*, **47**, 3428 (1982); *Tetrahedron Lett.*, 3311 (1979).
88. Ahmed, Z., and Cava, M. P., *Tetrahedron Lett.*, **22**, 5239 (1981).
89. Danishefsky, S., and Feng, Y. C., *Synth. Comm.*, **8**, 211 (1978).
90. Ryn, I., Murai, S., Niwa, I., and Sonoda, N., *Synthesis*, 874 (1977),
91. (a) Danishefsky, S., Yan, C.-F., Singh, R. K., Gammill, R. B., McCurry, P. M., Jr., Fritsch, N., and Clardy, J., *J. Am. Chem. Soc.*, **101**, 7001 (1979); (b) Danishefsky, S., Yan, C. F., and McCurry, P. M., Jr., *J. Org. Chem.*, **42**, 1819 (1977).

92. Danishefsky, S., Zamboni, R., Kahn, M., and Etheredge, S. J., *J. Am. Chem. Soc.,* **103**, 3460 (1981); ibid , **102**, 2097 (1980).
93. Semmelhack, M. F., and Zask, A., *J. Am. Chem. Soc.,* **105**, 2034 (1983).
94. Tietze, L. F., Kiedrowski, G., and Berger, B., *Tetrahedron Lett.,* **23**, 51 (1982).
95. Goldsmith, D. J., and Kezar, H. S., III, *Tetrahedron Lett.,* **21**, 3543 (1980).
96. Schlessinger, R. H., and Roberts, M. R., *J. Am. Chem. Soc.,* **103**, 724 (1981).
97. Nicolaou, K. C., Magolda, R. L., and Sipio, W. J., *Synthesis,* 982 (1979).
98. Nicolaou, K. C., Magolda, R. L., and Claremon, D. A., *J. Am. Chem. Soc.,* **102**, 1404 (1980).
99. Herron, J. N., and Pinder, A. R., *J. C. S. Perkin I,* 161 (1983).
100. Ohfune, Y.,and Tomita, M., *J. Am. Chem. Soc.,* **104**, 3511 (1982).
101. Williams, D. R., and Nishitani, K., *Tetrahedron Lett.,* **21**, 4417 (1980).
102. Williams, D. R., Barner, B. A., Nishitani, K., and Philips, J. G., *J. Am. Chem. Soc.,* **104**, 4708 (1982).
103. Crouse, G. D., and Paquette, L. A., *J. Org. Chem.,* **46**, 4272 (1981).
104. Boeckman, R. K., Jr., and Ko, S. S., *J. Am. Chem. Soc.,* **102**, 7146 (1980).
105. Brocksom, T. J., Petragnani, N., and Rodriguez, R., *J. Org. Chem.,* **39**, 2114 (1974).
106. Buddhsukh, D., and Magnus, P., *J. C. S. Chem. Comm.,* 952 (1975).
107. Kocienski, P. J., Cernigliaro, G., and Feldstein, G., *J. Org. Chem.,* **42**, 353 (1977).
108. Danishefsky, S., Hirama, M., Gombatz, K., Harayama, T., Berman, E., and Schuda, P. F., *J. Am. Chem. Soc.,* **101**, 7020 (1979); *ibid,* **100**, 6536 (1978).
109. Overman, C. E., and Fukaya, C., *J. Am. Chem. Soc.,* **102**, 1454 (1980).
110. (a) Cossey, A. L., Lombardo, L., and Mander, L., *Tetrahedron Lett.,* **21**, 4383 (1980); (b) For a related application of the reaction, see: Lombardo, and Mander, L., *J. Org. Chem.,* **48**, 2298 (1983).
111. Oppolzer, W., and Thirring, K., *J. Am. Chem. Soc.,* **104**, 4978 (1982).
112. Robins, D. J., and Sakdara, S., *J. C. S. Perkin I,* 1734 (1979).
113. Terao, Y., Imai, N., Achiwa, K., and Sekiya, M., *Chem. Pharm. Bull.,* **30**, 3167 (1982).
114. Vedejs, E., and Martinez, G. R., *J. Am. Chem. Soc.,* **102**, 7993 (1980).
115. Vedjes, E., Post-ICOS-IV Kyoto Symposium on Perspectives in Organic Synthesis, *Proceedings*, 1982, p. 59.
116. Wilson, C. A., II, and Bryson, T. A., *J. Org. Chem.,* **40**, 800 (1975).

117. Trost, B. M., Salzmann, T. N., and Hiroi, K., *J. Am. Chem. Soc.,* **98**, 4887 (1976). See also ref. 128.
118. Rao, Y. S., *Chem. Rev.,* **76**, 625 (1976); *ibid*, **64**, 353 (1964).
119. Marshall, J. A., and Ellison, R. H., *J. Am. Chem. Soc.,* **98**, 4312 (1976).
120. Ohloff, G., Giersch, W., Schulte-Elte, K. H., and Vial, C., *Helv. Chim. Acta,* **59**, 1140 (1976).
121. Wakamatsu, T., Akasaka, K., and Ban, Y., *J. Org. Chem.,* **44**, 2008 (1979); *Tetrahedron Lett.,* 2755 (1977).
122. Tomioka, K., Ishiguro, T., and Koga, K., *J. Chem. Soc. Chem. Comm.,* 652 (1979); *Tetrahedron Lett.,* **21**, 2973 (1980).
123. Vigneron, J. P., Méric, R., Larcheveque, M., Debal, A., Kunesch, G., Zaggatti, P., and Gallois, M., *Tetrahedron Lett.,* **23**, 5051 (1982).
124. White, J. D., Matsui, T., and Thomas, J. A., *J. Org. Chem.,* **46**, 3376 (1981).
125. Grieco, P. A., Pogonowski, C. S., and Burke, S., *J. Org. Chem.,* **40**, 542 (1975).
126. Semmelhack, M. F., and Tomoda, S., *J. Am. Chem. Soc.,* **103**, 2427 (1981). See also ref. 84a.
127. Smith, A. B., III, and Richmond, R. E., *J. Org. Chem.,* **46**, 4814 (1981).
128. Smith, A. B., III, and Richmond, R. E., *J. Am. Chem. Soc.,* **105**, 575 (1983).
129. Kido, F., Nada, Y., and Yoshikoshi, A., *J. C. S. Chem. Comm.,* 1209 (1982).
130. Luteijn, J. M., Doorn, M., and Groot, A., *Tetrahedron Lett.,* **21**, 4127 (1980).
131. Shieh, H.-M., and Prestwich, G. D., *Tetrahedron Lett.,* **23**, 4643 (1982).
132. Reviews: (a) Grieco, P. A., *Synthesis*, 67 (1975); (b) Gammill, R. B., Wilson, C. A., and Bryson, T. A., *Synth. Comm.,* **5**, 245 (1975); (c) Newaz, S. S., *Aldrichimica Acta,* **10**, 64 (1977).
133. Grieco, P. A., and Miyashita, M., *J. Org. Chem.,* **39**, 120 (1974).
134. Yamakawa, K., Nishitani, K., and Tominaga, T., *Tetrahedron Lett.,* 2829 (1975).
135. Grieco, P. A., and Nishizawa, M., *J. C. S. Chem. Comm.,* 582 (1976).
136. Yamakawa, K., Tominaga, T., and Nishitani, K., *Tetrahedron Lett.,* 4137 (1975).
137. Grieco, P. A., and Nishizawa, M., *J. Org. Chem.,* **42**, 1717 (1977).
138. Niwa, H., Kuroda, A., and Yamada, K., *Chem. Lett.,* 125 (1983).
139. Zoretic, P. A., and Soja, P., *J. Org. Chem.,* **41**, 3587 (1976); *J. Heterocyclic Chem.,* **14**, 681 (1977).
140. Ho, K., Suzuki, F., and Haruna, M., *J. C. S. Chem. Comm.,* 733 (1978).
141. Kozikowski, A. P., and Arnes, A., *J. Am. Chem. Soc.,* **103**, 3923 (1981).

142. Shinmon, N., Cava, M. P., and Brown, R. F. C., *J. C. S. Chem. Comm.*, 1020 (1980).
143. Brattesani, D. N., and Heathcock, C. H., *Tetrahedron Lett.*, 2279 (1974).
144. Brattesani, D. N., and Heathcock, C. H., *J. Org. Chem.*, **40**, 2165 (1975).
145. Grieco, P. A., and Yokoyama, Y., *J. Am. Chem. Soc.*, **99**, 5210 (1977).
146. Renga, J. M., and Reich, H. J., *Org. Syn.*, **59**, 58.
147. Marx, J. N., and Minaskanian, G., *Tetrahedron Lett.*, 4175 (1979); *J. Org. Chem.*, **47**, 3306 (1982).
148. Welch, S. C., Hagan, C. P., White, D. H., Fleming, W. P., and Trotter, J. W., *J. Am. Chem. Soc.*, **99**, 549 (1977).
149. Liotta, D., Barnum, C., Puleo, R., Zima, G., Bayer, C., and Kezar, H. S., III, *J. Org. Chem.*, **46**, 2920 (1981).
150. Liotta, D., Saindane, M., Barnum, C., Ensley, H., and Balakrishnan, P., *Tetrahedron Lett.*, **22**, 3043 (1981).
151. Ishida, T., and Wada, K., *J. C. S. Chem. Comm.*, 337 (1977).
152. Isobe, K., Taga, J., and Tsuda, Y., *Tetrahedron Lett.*, 2331 (1976).
153. Kozikowski, A. P., Schmiesing, R. J., and Sorgi, K. C., *J. Am. Chem. Soc.*, **102**, 6577 (1980).
154. Isobe, M., Kitamura, M., and Goto, T., *J. Am. Chem. Soc.*, **104**, 4997 (1982); *Tetrahedron Lett.*, **22**, 239 (1981).
155. (a) Pitteloud, R., and Petrzilka, M., *Helv. Chim. Acta*, **62**, 1319 (1979); (b) Petrzilka, M., *Helv. Chim. Acta*, **61**, 2286 (1978).
156. Metz, P., and Shafer, H. J., *Tetrahedron Lett.*, **23**, 4067 (1982).
157. (a) Hsu, C.-T., Wang, N.-Y., Latimer, L. H., and Sih, C. J., *J. Am. Chem. Soc.*, **105**, 593 (1983); (b) Wang, N.-Y., Hsu, C.-T., and Sih, C. J., *J. Am. Chem. Soc.*, **103**, 6538 (1981).
158. Arunachalam, T., and Caspi, E., *J. Org. Chem.*, **46**, 3415 (1981).
159. Liu, P. S., Marguez, V. E., Kelly, J. A., and Driscoll, J. S., *J. Org. Chem.*, **45**, 5227 (1980).
160. Beau, J. M., Aburaki, S., Pugny, J. R., and Sinay, P., *J. Am. Chem. Soc.*, **105**, 621 (1983).
161. Ohta, S., and Kimoto, S., *Tetrahedron Lett.*, 2279 (1975).
162. Tsuda, Y., Sano, T., Taka, J., Isobe, K., Toda, J., Irie, H., Tanaka, H., Takagi, S., Yamaki, M., and Murata, M., *J. C. S. Chem. Comm.*, 933 (1975).
163. Schlessinger, R. H., and Lopes, A., *J. Org. Chem.*, **46**, 5252 (1981).
164. (a) McGowan, D. A., and Berchtold, G. A., *J. Am. Chem. Soc.*, **104**, 1153 (1982); (b) Hoare, J. H., Policastro, P. P., and Berchtold, G. A., *J. Am.*

Chem. Soc., **105**, 6264 (1983).

165. Still, W. C., Murata, S., Revial, G., and Yoshihara, K., *J. Am. Chem. Soc.*, **105**, 625 (1983).
166. Trost, B. M., and Scudder, P. H., *J. Am. Chem. Soc.*, **99**, 7601 (1977).
167. Trost, B. M., Nishimura, Y., and Yamamoto, K., *J. Am. Chem. Soc.*, **101**, 1328 (1979).
168. Grieco, P. A., Noguer, J. A., and Masaki, Y., *Tetrahedron Lett.*, 4213 (1975).
169. Clark, R. D., and Heathcock, C. H., *J. Org. Chem.*, **41**, 1396 (1976).
170. Isobe, M., Iio, H., Kawai, T., and Goto, T., *J. Am. Chem. Soc.*, **100**, 1940 (1978); *Tetrahedron Lett.*, 703 (1977).
171. Bentley, P. H., and Hunt, E., *J. C. S. Chem. Comm.*, 518 (1978).
172. Kozikowski, A. P., and Chen, Y. Y., *J. Org. Chem.*, **46**, 5248 (1981).
173. Matsuda, F., Tomiyoshi, N., Yanagiya, M., and Matsumoto, T., *Tetrahedron Lett.*, **24**, 1277 (1983).
174. Grieco, P. A., Masai, Y., and Boyler, D., *J. Am. Chem. Soc.*, **97**, 1597 (1975).
175. Rueger, H., and Benn, M. H., *Can. J. Chem.*, **60**, 2918 (1982).
176. Kametani, T., Matsumoto, H., Nemoto, H., and Fukumoto, K., *J. Am. Chem. Soc.*, **100**, 6218 (1978).
177. Still, W. C., *J. Am. Chem. Soc.*, **101**, 2493 (1979).
178. Stork, G., and Morgans, D. J., Jr., *J. Am. Chem. Soc.*, **101**, 7110 (1979).
179. Grieco, P. A., Williams, E., Tanaka, H., and Gilman, S., *J. Org. Chem.*, **45**, 3537 (1980).
180. Morgans, D. J., Jr., *Tetrahedron Lett.*, **22**, 3721 (1981).
181. Snowden, R. L., *Tetrahedron Lett.*, **22**, 101 (1981).
182. Funk, R. L., Horcher, L. H. M., II, Daggett, J. U., and Hansen, M. M., *J. Org. Chem.*, **48**, 2632 (1983).
183. Oppolzer, W., Battig, K., and Hudlicky, T., *Helv. Chim. Acta*, **62**, 1493 (1979).
184. Woodward, R. B., *et. al.*, *J. Am. Chem. Soc.*, **103**, 3210 (1982).
185. Nishizawa, M., Grieco, P. A., Burke, S. D., and Metz, W., *J. C. S. Chem. Comm.*, 76 (1978).
186. Arno, M., Garcia, B., Pedro, J. R., and Seoane, E., *Tetrahedron Lett.*, **24**, 1741 (1983).
187. Grieco, P. A., Takigawa, T., and Schillinger, W. J., *J. Org. Chem.*, **45**, 2247 (1980).
188. Roush, W. R., and D'Ambra, T. E., *J. Org. Chem.*, **46**, 5045 (1981).
189. Takano, S., Takahashi, M., and Ogasawara, K., *J. Am. Chem. Soc.*, **102**, 4282 (1980).

190. (a) Ley, S. V., Simpkins, N. S., and Whittle, A. J., *J. C. S. Chem. Comm.*, 503 (1983); (b) Ley, S. V., Neuhaus, D., Simpkins, N. S., and Whittle, A. J., *J. C. S. Perkin I,* 2157 (1982); (c) Ley, S. V., Simpkins, N. S., and Whittle, A. J., *J. C. S. Chem. Comm.,* 1001 (1981).
191. Scarborough, R. M., Jr., Toder, B. H., and Smith, A. B., III, *J. Am. Chem. Soc.,* **102**, 3904 (1980).
192. Hoye, T. R., and Caruso, A. J., *Tetrahedron Lett.,* 4611 (1978).
193. Goldsmith, D. J., and Thottathil, I. K., *Tetrahedron Lett.,* **22**, 2447 (1981).
194. Takano, S., Tamura, N., and Ogasawara, K., *J. C. S. Chem. Comm.,* 1155 (1981).
195. Takano, S., Tamuro, N., Ogasawara, K., Nakagawa, Y., and Sakai, T., *Chem. Lett.*, 933 (1982).
196. Arita, M., Adachi, K., Ito, Y., Sawai, H., and Ohno, M., *J. Am. Chem. Soc.,* **105**, 4049 (1983).
197. Kometani, T., Takeuchi, Y., and Yoshii, E., *J. Org. Chem.,* **48**, 2630 (1983).
198. Mori, K., *Tetrahedron,* **32**, 1101 (1976).
199. Masamune, S., Kaiho, T., and Garvey, D. S., *J. Am. Chem. Soc.,* **104**, 5521 (1982).
200. Heck, J. V., and Christensen, B. G., *Tetrahedron Lett.,* **22**, 5027 (1981).
201. Sharpless, K. B., and Lauer, R. F., *J. Am. Chem. Soc.,* **94**, 7154 (1972).
202. Reich, H. J., *J. Org. Chem.,* **40**, 2570 (1975).
203. Clive, D. L. J., *J. C. S. Chem. Comm.,* 771 (1978).
204. Salmond, W. G., Barta, M. A., Cain, A. M., and Sobala, M. C., *Tetrahedron Lett.,* 1683 (1977).
205. Wollenberg, R. H., *Tetrahedron Lett.,* **21**, 3139 (1980).
206. Reich, H. J., and Shah, S. K., *J. Am. Chem. Soc.,* **99**, 263 (1977).
207. Reich, H. J., Shah, S. K., Gold, P. M., and Olson, R. E., *J. Am. Chem. Soc.,* **103**, 3112 (1981).
208. (a) Reich, H. J., and Wollowitz, S., *J. Am. Chem. Soc.,* **104**, 7051 (1982); (b) Reich, H. J., Reich, I. L., and Wollowitz, S., *J. Am. Chem. Soc.,* **100**, 5981 (1978); (c) Reich, I. L., and Reich, H. J., *J. Org. Chem.,* **46**, 3721 (1981).
209. Nishiyama, H., Hagagi, K., Sakuta, K., and Hoh, K., *Tetrahedron Lett.,* **22**, 5285 (1983).
210. Nishiyama, H., Narimatsu, S., and Hoh, K., *Tetrahedron Lett.,* **22**, 5289 (1981).
211. Hutchinson, C. R., Post-ICOS-IV Kyoto Symposium on Perspectives in

Organic Synthesis, Proceedings, p. 41, 1982.
212. Magnus, P., Cooke, F., and Sarkar, T., *Organometallics,* **1**, 562 (1982).
213. Zimmerman, H. E., and Diehl, D. R., *J. Am. Chem. Soc.,* **101**, 1841 (1979).
214. Smith, A. B., III, Levenberg, P. A., Jerris, P. J., Scarborough, R. M., Jr., and Wovkulich, P. M., *J. Am. Chem. Soc.,* **103**, 1501 (1981).
215. Reich, H. J., and Chow, F., *J. C. S. Chem. Comm.,* 1136 (1975).
216. (a) Rémion, J., Durmont, W., and Krief, A., *Tetrahedron Lett.,* 1385 (1976); (b) Rémion, J., and Krief, A., *Tetrahedron Lett.,* 3743 (1976).
217. Halazy, S., and Krief, A., *J. C. S. Chem. Comm.,* 1136 (1979).
218. Nicolaou, K. C., Sipio, W. J., Magolda, R. L., and Claremon, D. A., *J. C. S. Chem. Comm.,* 83 (1979).
219. Clive, D. L. J., and Kalé, V. N., *J. Org. Chem.,* **46**, 231 (1981).
220. Léonard-Coppens, A. M., and Krief, A., *Tetrahedron Lett.,* 3227 (1976).
221. Sevrin, M., Denis, J. N., and Krief, A., *Tetrahedron Lett.,* **21**, 1877 (1980).
222. Labar, D., Hevesi, L., Dumont, W., and Krief, A., *Tetrahedron Lett.,* 1141 (1978).
223. Miyoshi, N., Yamamoto, T., Kambe, N., Murai, S., and Sonoda, N., *Tetrahedron Lett.,* **23**, 4813 (1982).
224. Jefson, M., and Meinwald, J., *Tetrahedron Lett.,* **22**, 3561 (1981).
225. Hayakawa, H., Tanaka, H., and Miyasaka, T., *Chem. Pharm. Bull.,* **30**, 4589 (1982).
226. Kutney, J. P. and Singh, A. K., *Can. J. Chem.,* **61**, 1111 (1983).
227. Exon, C., Gallagher, T., and Magnus, P., *J. Am. Chem. Soc.,* **105**, 4739 (1983).
228. Kuwajima, I. And Shimizu, M., *Tetrahedron Lett.,* 1277 (1978).
229. Schwartz, J., and Hayashi, Y., *Tetrahedron Lett.,* **21**, 1497 (1980).
230. Ando, M., Yamaoka, H., and Takase, K., *Chem. Lett.,* 501 (1982).
231. Nakatsuka, S., Yamada, K., Yoshida, K., Asano, O., Murakami, Y., and Goto, T., *Tetrahedron Lett.,* **24**, 5627 (1983);
232. Takano, S., Morimoto, M., Masuda, K., Ogasawara, K., *Chem. Pharm. Bull.,* **30**, 4238 (1982).
233. Sirat, H. M., Thomas, E. J., and Wallis, J. D., *J. C. S. Perkin I,* 2885 (1982).

CHAPTER 5

Organoselenium-mediated Reductions

The wide application of selenides for the introduction of olefinic bonds has resulted in the development of numerous methods for their synthesis and oxidative removal (Chap. 4 and Chap. 6). Some non-oxidative transformations of these compounds, however, were also developed, such as: reductive cleavage to replace the C–Se bond with a C–H bond, and conversion of β-hydroxyselenides to epoxides. These reactions, together with some reductions induced by benzeneselenol (PhSeH), are the subject of this chapter.

5.1 Reductive elimination of selenides

The reductive cleavage of aryl (or methyl) alkyl selenides (**1**), *i.e.* substitution of the SeAr (or SeMe) group with an H (**1**→ **2**, Eq. 5.1), has been achieved by various methods (Table 5.1).

$$\underset{\mathbf{1}\ (R'=\mathrm{Ar,\ Me})}{\mathrm{RSeR'}} \xrightarrow{[\mathrm{H}]} \underset{\mathbf{2}}{\mathrm{RH}} \qquad (5.1)$$

Table 5.1. Methods for the reductive cleavage of selenides.

Reagent	*Ref.*
Li–$EtNH_2$	1
Raney Ni	1–5
Ph_3SnH	6, 7
n-Bu_3SnH, AIBN	4–8
$Pd(Ph_3P)_4$, $LiBHEt_3$	9
P_2I_4	10

Some illustrative examples of this type of reduction are given in Eq. 5.2,[1] 5.3,[2,4] 5.4,[7] 5.5,[5] 5.6[9] and 5.7.[10]

Li–EtNH$_2$, −10° C (77%) (5.2)

Raney Ni (94%) (5.3)

Ph_3SnH, PhMe, 120° C (88%) (5.4)

n-Bu_3SnH, AIBN, PhMe, 110° C (5.5)

$Pd(Ph_3P)_4$, $LiBHEt_3$ (82%) (5.6)

P_2I_4 (95%) (5.7)

When combined with the facile formation of selenides from primary alcohols (§4.4), the reaction is an efficient process for the selective overall deoxygenation of primary alcohols. An example is shown in Eq. 5.8, taken from Grieco's approach to the macrolide tylosin.[11]

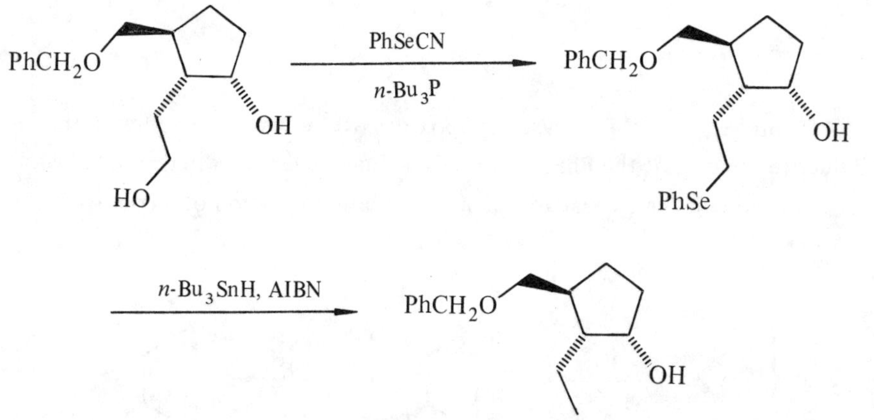
PhCH₂O
OH
HO
PhSeCN
n-Bu₃P
PhCH₂O
OH
PhSe
n-Bu₃SnH, AIBN
PhCH₂O
OH

Clive[12,13] and Krief[14–17] have developed methods for the conversion of ketones and aldehydes (3) to the corresponding selenoketals and selenoacetals (4). Combined with the reductive removal of the selenide groups, this process constitutes an overall deoxygenation of carbonyl compounds (Eq. 5.9). For example, cholestanone (6) is efficiently converted to cholestane (7, Eq. 5.10).

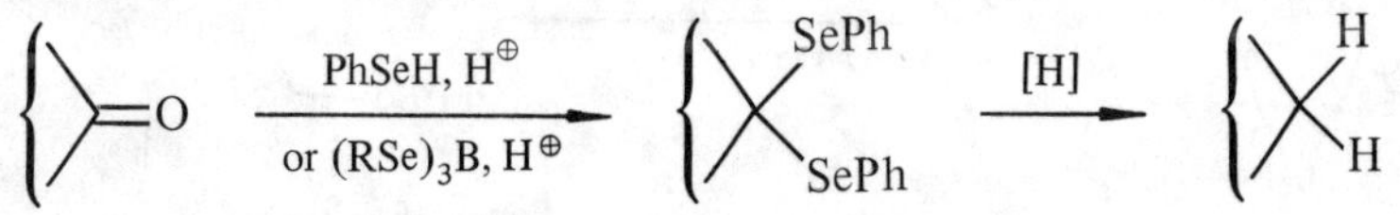
O
PhSeH, H⊕
or (RSe)₃B, H⊕
SePh
SePh
[H]
H
H

(5.9)

3 4 5

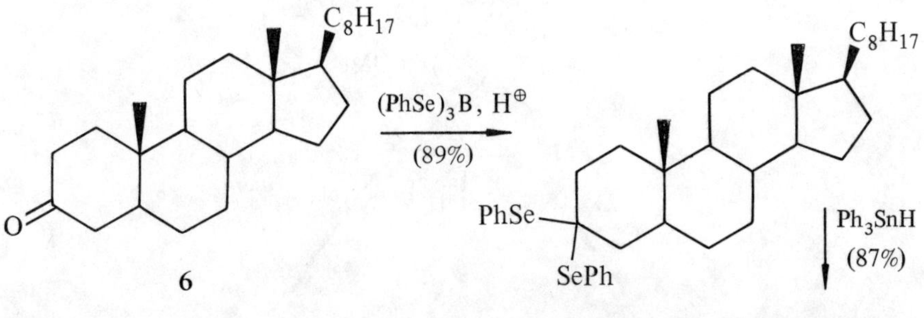
C₈H₁₇
O
(PhSe)₃B, H⊕
(89%)
C₈H₁₇
PhSe
SePh
Ph₃SnH
(87%)

6

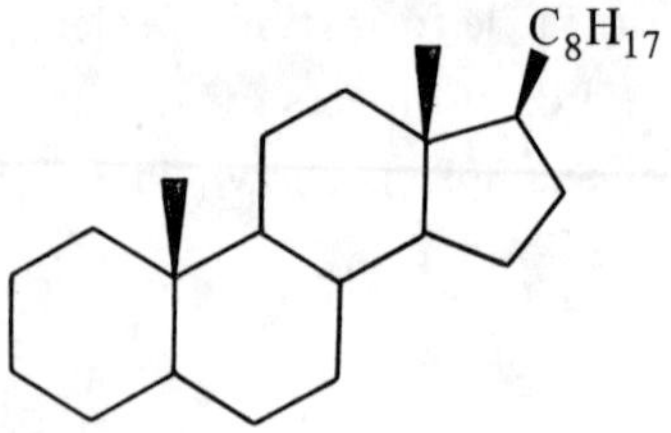

C_8H_{17}

7

(5.10)

Combination of the cleavage of epoxides (**8**) by $PhSe^{\ominus}$ (§4.3) with the reductive removal of the PhSe group, results in an overall reductive opening of epoxides, with regioselective and stereospecific formation of alcohols (**9**, Eq. 5.11).

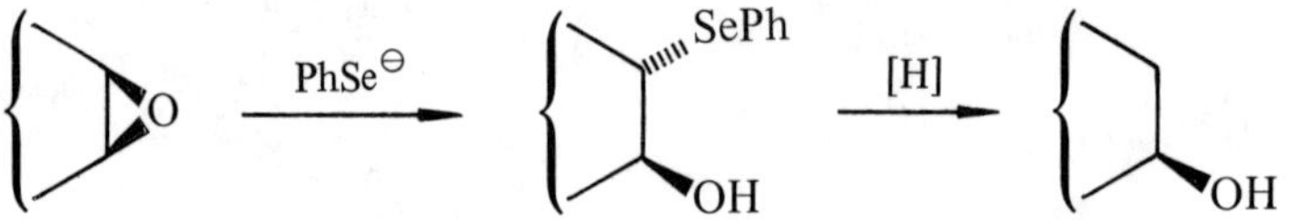

$PhSe^{\ominus}$ SePh OH [H] OH (5.11)

8 **9**

The method was used by Joullié in the synthesis of muscarine analogs such as D-isoepiallomuscarine (**10**),[18] and in the synthesis of the antibiotic furanomycin (**11**),[19] as shown in Eq. 5.12.

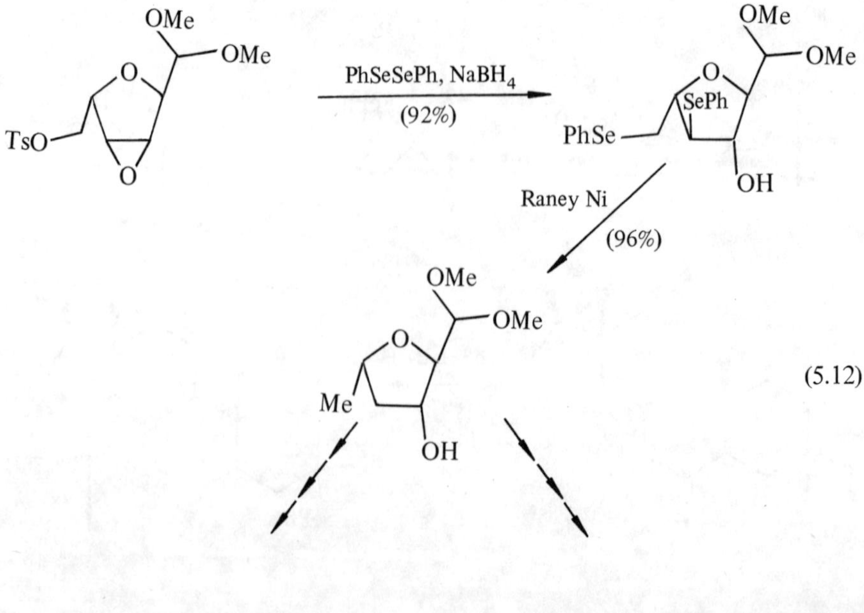

OMe OMe O TsO O PhSeSePh, $NaBH_4$ (92%) OMe OMe O SePh PhSe OH Raney Ni (96%) OMe OMe O Me OH

(5.12)

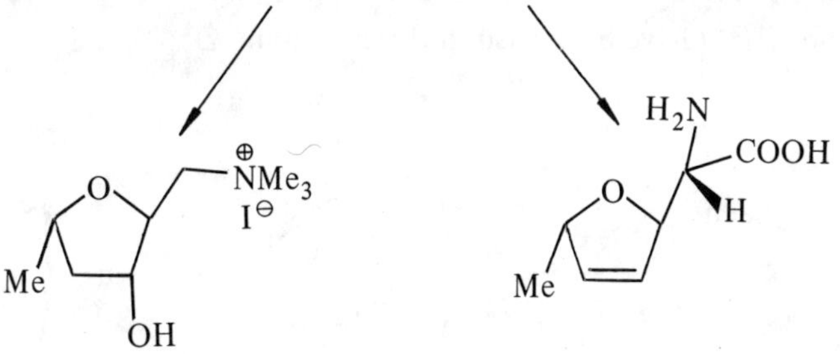

10

11

(5.12)

Using an alkoxyselenation type reaction (§4.2), followed by reductive removal of the PhSe group, Sinay has developed a novel synthesis of 2′-deoxydisaccharides such as **12** (Eq. 5.13).

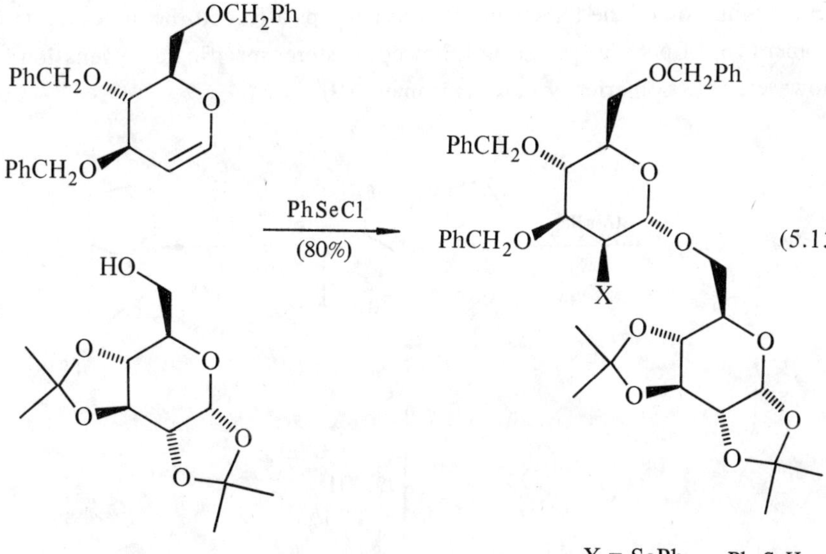

PhSeCl (80%)

X = SePh → Ph_3SnH (95%) → **12**, X = H

(5.13)

5.2 Conversion of β-hydroxyselenides to epoxides

Alkylation of β-hydroxyselenides (**13**) at selenium and subsequent treatment of the resulting β-hydroxyselenonium salts (**14**) with base, affords the epoxides (**15**, Eq. 5.14). The process is the opposite of the nucleophilic opening of epoxides with selenide anion (Eq. 4.111). Several alkylative agents ($Et_3O^{\oplus}BF_4^{\ominus}$,[21]

MeI–$AgBF_4$,[22] MeI,[23, 24] $MeSO_3F$[17]) and several bases (NaH,[21] *t*-BuOK,[22–24] KOH[17] and KH[25]) have been used in this reaction.

SePh, OH, 13; RX; Ph, ⊕Se, $X^{\ominus}$, R, OH, 14; Base; O, 15

(5.14)

An application of the method occurs in Paquette's synthesis of the pheromone multifidene (**19**), shown in Eq. 5.15.[25] β-Hydroxyselenide **17**, prepared by addition of ethyl magnesium bromide to aldehyde **16** (Eq. 4.37), upon direct elimination of the PhSe and hydroxyl groups afforded the undesired *trans* isomer (**18**). Epoxide formation followed by stereospecific deoxygenation, however, gave isomerically pure *cis* isomer (**19**).

CHO, SePh, **16**; EtMgBr (76%); OH, SePh, **17**; $SOCl_2$, Et_3N (64%); **18**

(1) $Et_3O^{\oplus}BF_4^{\ominus}$ (2) KH (71%)

O, H, H; (1) Ph_2PI (2) MeI (51%); **19**

(5.15)

5.3 Reductions induced by benzeneselenol

In the presence of base, heat or light, benzeneselenol (PhSeH) acts as a reducing agent for a variety of functional groups, such as nitro compounds,[26] azo compounds,[26,27] hydrazones,[28] diazonium salts,[29] sulphoxides,[27,30] *α,β*-unsaturated carbonyl compounds[28] and iodomethyl ketones.[31] Typical examples are given in Eq. 5.16,[26] 5.17,[28] 5.18,[29] 5.19,[30] 5.20[28] and 5.21.[31]

$$RNO_2 \text{ or } RN{=}NR \xrightarrow[\text{DABCO}\ (100\%)]{\text{PhSeH}} RNH_2 \quad (5.16)$$

$$\text{PhCH=N–NHPh} \xrightarrow[(95\%)]{\text{PhSeH}, h\nu} \text{PhCH}_2\text{NHNHPh} \quad (5.17)$$

$$\xrightarrow[(70\%)]{\text{PhSeH}, \Delta} \quad (5.18)$$

$$\xrightarrow[(84\%)]{\text{PhSeH}, \Delta} \quad (5.19)$$

$$\text{PhCH=CHCOPh} \xrightarrow[(95\%)]{\text{PhSeH}, h\nu} \text{PhCH}_2\text{CH}_2\text{COPh} \quad (5.20)$$

$$\text{RCOCH}_2\text{I} \xrightarrow[(100\%)]{\text{PhSeH}, K_2CO_3} \text{RCOMe} \quad (5.21)$$

REFERENCES

1. Sevrin, M., Van Ende, D., and Krief, A., *Tetrahedron Lett.*, 2643 (1976).
2. Nicolaou, K. C., and Lysenko, Z., *Tetrahedron Lett.*, 1257 (1977).
3. Nicolaou, K. C., and Lysenko, Z., *J. Am. Chem. Soc.*, **99**, 3185 (1977).
4. Nicolaou, K. C., Magolda, R. L., Sipio, W. J., Barnette, W. E., Lysenko, Z., and Joullié, M. M., *J. Am. Chem. Soc.*, **102**, 3784 (1980).
5. Nicolaou, K. C., Seitz, S. P., Sipio, W. J., and Blount, J. F., *J. Am. Chem. Soc.*, **101**, 3884 (1979).
6. Clive, D. L. J., Chittattu, G. J., and Wong, C. K., *J. C. S. Chem. Comm.*, 41 (1978).
7. Clive, D. L. J., Chittattu, G. J., Farina, V., Kiel, W. A., Menchen, S. M., Russell, C. G., Singh, A., Wong, C. K., and Curtis, J. J., *J. Am. Chem. Soc.*, **102**, 4438 (1980).
8. Corey, E. J., Pearce, H. L., Székely, I., and Ishiguro, M., *Tetrahedron Lett.*, 1023 (1978).
9. Hutchins, R. O., and Learn, K., *J. Org. Chem.*, **47**, 4380 (1982).
10. Denis, J. N., and Krief, A., *J. C. S. Chem. Comm.*, 229 (1983).
11. Grieco, P. A., Inanaga, J., Lin, N.-H., and Yanami, T., *J. Am. Chem. Soc.*, **104**, 5781 (1982).
12. Clive, D. L. J., and Menchen, S. M., *J. Org. Chem.*, **44**, 4279 (1979); *J. C. S. Chem. Comm.*, 356 (1978).
13. Clive, D. L. J., and Menchen, S. M., *J. Org. Chem.*, **44**, 1883 (1979).
14. Dumont, W., and Krief, A., *Angew. Chem. Int. Ed. Engl.*, **16**, 540 (1977).
15. Dumont, W., Sevrin, M., and Krief, A., *Angew. Chem. Int. Ed. Engl.*, **16**, 541 (1977).
16. Sevrin, M., and Krief, A., *Tetrahedron Lett.*, 187 (1978).
17. Krief, A., *Tetrahedron*, **36**, 2531 (1980).
18. (a) Wang, P. C., and Joullié, M. M., *J. Org. Chem.*, **45**, 5359 (1980); (b) Wang, P. C., Lysenko, Z., and Joullié, M. M., *Heterocycles*, **9**, 753 (1978).
19. Semple, J. E., Wang, P. C., Lysenko, Z., and Joullié, M. M., *J. Am. Chem. Soc.*, **102**, 7505 (1980).
20. Jaurand, G., Beau, J.-M., and Sinay, P., *J. C. S. Chem. Comm.*, 572 (1981).
21. Sharpless, K. B., Gordon, K. M., Lauer, R. F., Patrick, D. W., Singer, S. P., and Young, M. W., *Chem. Scr.*, **8A**, 9 (1975).
22. Dumont, W., and Krief, A., *Angew. Chem. Int. Ed. Engl.*, **14**, 350 (1975).

23. Van Ende, D., Dumont, W., and Krief, A., *Angew. Chem. Int. Ed. Engl.*, **14**, 700 (1975).
24. Van Ende, D., and Krief, A., *Tetrahedron Lett.*, 457 (1976).
25. Crouse, G. D., and Paquette, L. A., *J. Org. Chem.*, **46**, 4272 (1981).
26. Fujimori, K., Yoshimoto, H., and Oae, S., *Tetrahedron Lett.*, 4397 (1979).
27. Günther, W. H. H., *J. Org. Chem.*, **31**, 1202 (1966).
28. Perkins, M. J., Smith, B. V., and Turner, E. S., *J. C. S. Chem. Comm.*, 977 (1980).
29. James, F. G., Perkins, M. J., Porta, O., and Smith, B. V., *J. C. S. Chem. Comm.*, 131 (1977).
30. Perkins, M. J., Smith, B. V., Terem, B., and Turner, E. S., *J. Chem. Res. (S)*, 341 (1979).
31. Seshardi, R., Pegg, W. J., and Israel, M., *J. Org. Chem.*, **46**, 2596 (1981).

CHAPTER 6

Organoselenium-stabilized Carbanions

Due to its electronegative character, selenium has the useful property of stabilizing negative charge at a neighboring carbon. Selenium-stabilized carbanions are very versatile synthetic intermediates. They react with a variety of electrophiles, producing potentially useful compounds. Krief[1] and Reich,[2] among others, have studied and reviewed the chemistry of these anions. Their studies have turned up a number of new methods for the synthesis of olefins, alcohols, carbonyl compounds, epoxides and other types of compounds.

The most common selenium carbanions are those stabilized by a selenide group (PhSe or MeSe), and particularly those having additional stabilization by another functional group, such as carbonyl, nitrile, olefin, acetylene, phenylselenenyl and trimethylsilyl. The greater stability of the latter facilitates their preparation. Some typical applications of carbanions stabilized by selenides, selenoxides and selenonium salts (ylids) are described in this chapter.

Many of the selenium-stabilized carbanions can be considered as synthetic equivalents of various functionalities, summarized in Table 6.1.

Table 6.1. Synthetic equivalents of selenium stabilized carbanions.

Carbanion	*Synthetic equivalent*	
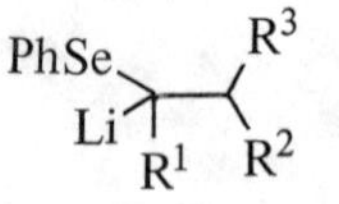 PhSe, Li, R^1, R^2, R^3	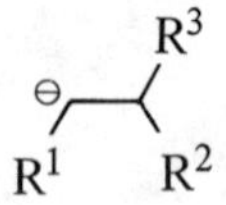 ⊖, R^1, R^2, R^3	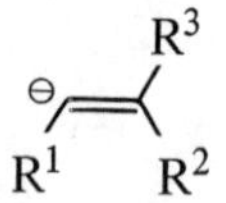 ⊖, R^1, R^2, R^3
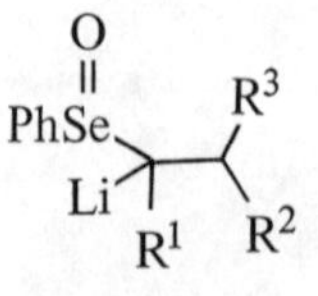 O, PhSe, Li, R^1, R^2, R^3	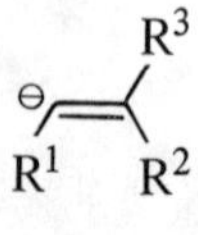 ⊖, R^1, R^2, R^3	

Table 6.1 (continued).

Carbanion	*Synthetic equivalent*	*Carbanion*	*Synthetic equivalent*
$PhSeC(Li)(COR^3)CHR^1R^2$	$R^1RC{=}C^{\ominus}H{-}COR^3$	$PhSeCH(Li){-}C{\equiv}C{-}Li$	$^{\ominus}CH{=}CH{-}CO^{\ominus}$
$PhSeCH(Li)CH{=}CH_2$	$^{\ominus}CH_2CH{=}CHCH_2OH$	$PhSeC(Li){=}C{=}CH_2$	$^{\ominus}CH_2C{\equiv}CCH_2OH$
$PhSeCH(Li)CH{=}C(Cl)R$	$^{\ominus}CH_2CH{=}CHCOR$	$PhSeC(Li){=}CR^1R^2$	$^{\ominus}COCHR^1R^2$; $H(^{\ominus})C{=}CR^1R^2$
$PhSeCH(Li)CH{=}CHSePh$	$^{\ominus}CH_2CH{=}CHCHO$	$(PhSe)_2C(Li)R$	$^{\ominus}COR$; $^{\ominus}CH_2R$
$PhSeCH(Li)CH{=}C(SePh)SiMe_3$	$^{\ominus}CH_2CH{=}CHCOSiMe_3$	$PhSeC(Li)(SiMe_3)R$	$^{\ominus}COR$
$PhSeCH_2C{\equiv}CLi$	$CH_2{=}C(SePh)CO^{\ominus}$	$PhSeCH(Li)SiMe_3$	$^{\ominus}CHO$

6.1 Reactions of the carbanions of alkyl selenides

α-Lithio selenides (**1**) can be prepared by a variety of methods, as summarized in Eq. 6.1. The most efficient and general method is the selenium–lithium exchange (**2**→**1**) of seleno ketals (**2**).[3–7] The latter can be synthesized from ketones (**3**) by ketalization with phenylselenol and acid catalyst[3,4,7–10] or with tris(phenylseleno)borane [$(PhSe)_3B$].[11] They can also be made by alkylation of anions derived from selenoacetals (**4**)[3,12–15] or selenoorthoesters (**5**).[3,14,15] The direct deprotonation of alkyl selenides (**6**) is not an efficient process,[3,16] because it requires an alkyllithium as the base, which leads to Se–Li exchange. However, when another anion-stabilizing group is also present, deprotonation can be effected efficiently with a lithium amide as the base (*e.g.* LDA),[16] as in the case of **4**→**2** (§6.2, §6.3). Vinyl selenides (**7**) can also be converted to **1**, either by a Michael-type addition of alkyl lithium,[17–19] or by converting them to *α*-bromoselenides (**8**) with hydrogen bromide,[9] which also give **1**, by Br–Li exchange.[9,16] Most of the above methods work not only with phenylselenides (Eq. 6.1), but also with the corresponding methylselenides.

A variety of electrophiles ($E^{\oplus}$) react with α-lithioselenides (**1**), leading to the corresponding α-substituted selenides (**9**, Eq. 6.1), as shown in Table 6.2.

Table 6.2. Addition of electrophiles to α-lithioselenides (**1**).

Electrophile ($E^{\oplus}$)	*E* (**9**)	*Ref.*
H_2O or D_2O	H or D	3, 19
Me_3SiCl	$SiMe_3$	14, 22, 23
PhSeBr	SePh	12, 22
PhSCl	SPh	4, 22
RX	R	12, 16, 19–21
Me_2NCHO (DMF)	CHO	6, 12, 19
RCOCl or $(RCO)_2O$	COR	6, 12, 19
$CO_2/H^{\oplus}$	COOH	6
ClCOOR	COOR	6
RCHO	CH(OH)R	4, 5, 24–27
R^1COR^2	$C(OH)R^1R^2$	3–5, 9, 24–30
$RCOCH{=}CR^1R^2$	$C(OH)RCH{=}CR^1R^2$	31
O, R^1, R^2	$CH_2C(OH)R^1R^2$	10, 12, 32
O, R^1, R^2	$CH_2CH_2C(OH)R^1R^2$	32

Further elaboration of selenides **9**, oxidatively (Chap. 4) or reductively (Chap. 5), gives various selenium-free products. For example, alkyl bromides (**10**) can be converted to terminal olefins (**13**) by treatment with $PhSeCH_2Li$ and subsequent removal of the PhSe group from the resulting selenide (**11**), either oxidatively or by conversion to the selenonium salt (**12**) and treatment with base (Eq. 6.2).[21]

$$\underset{\mathbf{10}}{R\frown Br} \xrightarrow[\text{THF–HMPT}]{PhSeCH_2Li} \underset{\mathbf{11}}{R\frown\frown SePh} \xrightarrow[\text{or MeI, AgBF}_4]{MeSO_3F}$$

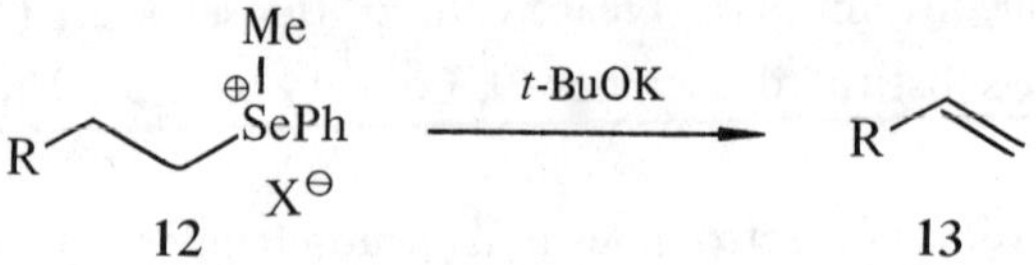

Me, ⊕SePh, X⊖ (12) —t-BuOK→ R (13) (6.2)

The reaction of selenide-stabilized carbanions (**1**) with carbonyl compounds (**14**) gives access to the synthetically useful β-hydroxyselenides (**15**). Intermediates of type **15** can therefore be obtained by combining two different carbonyl compounds (**3** and **14**), and they can be transformed into allylic alcohols (**16**, §4.3), α-selenocarbonyl compounds (**17**, §4.1), olefins (**18**, §4.6), epoxides (**19**, §5.2) or saturated alcohols (**20**, §5.1), as summarized in Eq. 6.3.

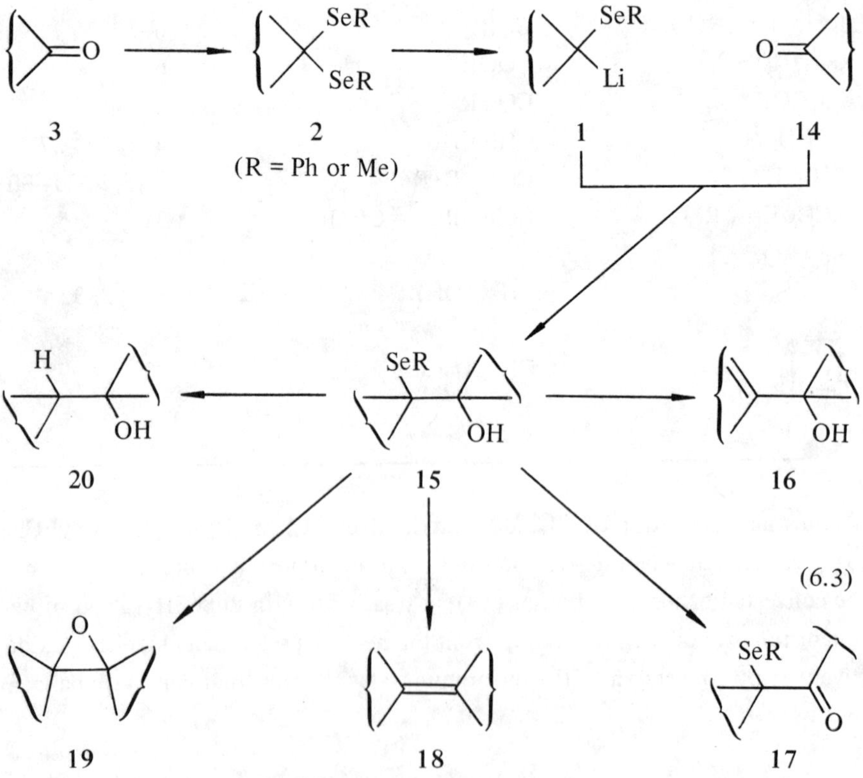

3 → 2 (R = Ph or Me) → 1 + 14 → 15 → 20, 16, 19, 18, 17 (6.3)

Application of the method to the synthesis of primary allylic alcohols (**21**) is exemplified in Eq. 6.4.[33] An alternative route to this useful class of compounds is given in Eq. 6.5.[34]

$PhSeH{-}H^{\oplus}$; (1) n-BuLi, (2) CH_2O (gas) (78%); t-BuOOH, Al_2O_3, 55° C (78%)

21

(6.4)

(1) n-BuLi, (2) DMF; $CH_2{=}PPh_3$; [O]

(6.5)

The overall conversion of carbonyl compounds (**14**) to olefins (**18**, Eq. 6.3) constitutes an equivalent synthesis to the Wittig reaction and has been used for the synthesis of sterically hindered olefins.[24–28, 35] A rather dramatic example is shown in Eq. 6.6.[28]

Et_2O, −78° C (83%); PI_3, NEt_3 (95%)

(6.6)

An interesting carbanion, thoroughly investigated by Krief, is α-selenolithiocyclopropane (**23**), which is derived from selenoketal **22** (Eq. 6.7).[15] As indicated in Eq. 6.7, conversion of **23** to cyclopropyl derivative **24** can be done by three alternative sequences (*A–C*), with quite different geometrical control.[27] Noteworthy is the conversion of either the *E*-isomer or the *Z*-isomer to the strained allylic alcohol **25**, via a [2,3]-sigmatropic rearrangement (§4.5).[36]

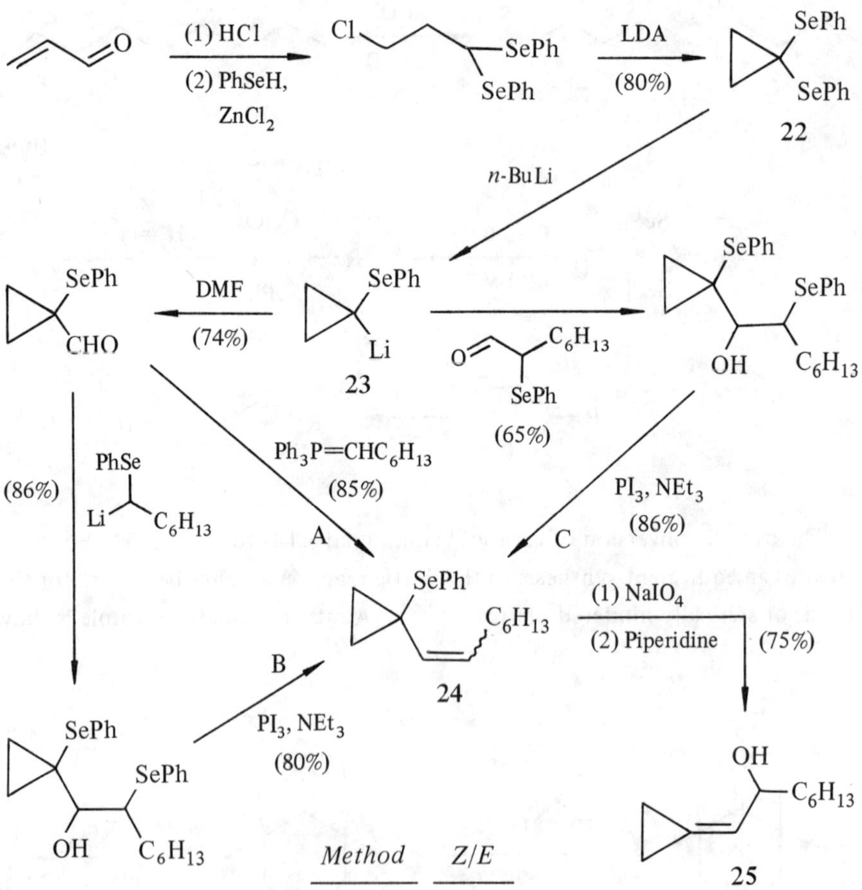

Method	*Z/E*
A	90/10
B	45/55
C	6/94

(6.7)

The similarly prepared methyl selenide **26** undergoes several interesting transformations, as shown in Eq. 6.8.[37]

SeMe

26

n-BuLi, CO_2 (65%)

COOH

P-TsOH (80%)

n-BuLi, MeI (77%)

Me

O

(6.8)

β-Hydroxyselenides derived by addition of **23** to ketones, form cyclobutanones on treatment with acid. The reaction parallels Trost's method with the corresponding phenylsulfide[38] and has been used by Krief in an efficient synthesis of α-cuparenone (**29**), outlined in Eq. 6.9.[39] Krief's synthesis also utilized a ring expansion of cyclobutanone **27** to cyclopentanone **29** via β-hydroxyselenide **28**,a process observed with several other adducts of hindered cyclic ketones with tertiary α-lithioselenides.[28–30]

O

Me

SePh Li

23 (81%)

HO SePh

Me

p-TsOH (70%)

O

Me 27

Li SeMe

Et_2O, −78° C (66%)

(6.9)

SeMe

OH

$AgBF_4$, Al_2O_3 (69%)
or EtOTl (57%)
or $MeOSO_2F$ (82%)

Me

29

Me

28

(6.9)

For the synthesis of the isomeric compound β-cuparenone (33), which is also naturally occurring, Krief used a slightly different approach, *i.e.* conversion of the similarly synthesized cyclobutanone **30** to β-hydroxyselenide **31**, formation of epoxide **32** and rearrangement to cyclopentanone **33**, as shown in Eq. 6.10.[39] The intermediate epoxide (**32**) was also converted selectively to the isomeric cyclopentanone (**35**) via the chlorohydrin (**34**, Eq. 6.10).[39]

O

SeMe
Li

OH

SeMe

Me

30

Me

31

EtOTl
(88%)

LiI, 12-C-4 (94%)

$BeCl_2$

$AgBF_4$, Al_2O_3 (75%)

Me 33 Me 32

OH Cl

Me 35 Me 34 (6.10)

The preparation of epoxides (**19**, Eq. 6.3) from β-hydroxyselenides (**15**) is a very useful process (§5.2), particularly when combined with the synthesis of **15** from selenide carbanions (**1**) and carbonyl compounds (**14**, Eq. 6.3). It can be applied efficiently to the synthesis of hindered epoxides[28, 29] as well as to the synthesis of α,β-unsaturated epoxides.[31] Krief has applied the method in a highly stereocontrolled synthesis of lanosterol side chain having the unnatural 20(R) stereochemistry (Eq. 6.11).[40] Interestingly, the reaction of epoxide **36** with ethoxyethynyl magnesium bromide proceeds with initial isomerization to aldehyde **37**, followed by addition to the aldehyde to form **38** (Eq. 6.11).[40]

(6.11)

The selective oxidation of β-hydroxyselenides (**15**, Eq. 6.3) derived from selenide carbanions (**1**) and aldehydes (**14**) to give the corresponding α-seleno carbonyl compounds (**17**) without oxidation of the selenide group, can be achieved with DDQ,[41] the Corey–Kim method (NCS–Me_2S–Et_3N),[41–43] the Posner method (Cl_3CCHO–Al_2O_3)[43, 44] or the Barton method (Ph_3BiCO_3).[45]

Additions of selenide carbanions (**1**) to epoxides (**39**) leads to γ-hydroxyselenides (**40**),[10, 12, 32] which can then be converted to homoallylic alcohols (**41**),[10] α,β-unsaturated carbonyl compounds (**42**),[10] γ-halohydrins (**43**),[1, 10] allylic alcohols (**44**)[1, 10] or oxetanes (**45**),[32] as indicated in Eq. 6.12.

SeR Li **1** + O **39**

SeR OH **40**

(1) MeI (2) t-BuOK or KOH or KH → **41** (OH)

Jones → **42** (O)

H_2O_2 → **44**

43 (X, OH) — MgO → **44** (OH)

43 — MeMgBr, t-BuOK → **45** (O)

43
X = Br (Br_2)
X = I (MeI–NaI, $CaCO_3$)

(6.12)

Krief has used this chemistry in an elegant synthesis of the pheromone ipsenol (**46**), shown in Eq. 6.13.[46]

O — MeSeH, $ZnCl_2$ (89%) → MeSe, SeMe — (1) n-BuLi (2) O-epoxide (73%) → MeSe, OH — (1) MeI, $AgBF_4$ (2) KH (77%) ↓

(6.13)

OH Δ (100%) OH

46 (6.13)

6.2 Reactions of the carbanions of α-phenylselenocarbonyl compounds.

α-Phenylselenocarbonyl compounds (**48**) readily obtained by selenenylation of the appropriate carbonyl precursors (**47**, §4.1), have enhanced acidity and can be converted to the corresponding carbanions with LDA. Reaction of these stabilized carbanions with electrophiles ($E^{\oplus}$) to give **49**, followed by oxidative elimination of the PhSe group, leads to α-substituted, α,β-unsaturated carbonyl compounds (**50**), as indicated in Eq. 6.14.

O (1) LDA (2) PhSeCl O SePh (1) LDA (2) $E^{\oplus}$

47 48

O E SePh [O] O E (6.14)

49 50

Grieco has applied the method to the conversion of **51** to **52**, which was elaborated to give the key intermediate **55**, used in his synthesis of vernolepin (**194**, Eq. 4.123) and vernomenin (**56**), as shown in Eq. 6.15.[47] The synthesis also included the conversion of alcohol **53** to terminal olefin **54** via organoselenium chemistry (§4.4). A sequence similar to **51**→**52** was also carried out on intermediate **57** during Grieco's synthesis of temisin (**213**, Eq. 4.140).[48]

(1) LDA–PhSeCl
(2) LDA, HMPA, $BrCH_2CH{=}CMe_2$
(3) H_2O_2

X = H, Y = OMe : **51**
X = Me, Y = H : **57**

52

(6.15)

53

(1) $o\text{-}NO_2C_6H_4SeCN$, $n\text{-}Bu_3P$
or MsCl, ArSeNa
(2) H_2O_2

54

BBr_3

56

55

An efficient synthesis of **58** from cyclopentanone (**57**), was also based on this method (Eq. 6.16).[49] Compound **58** was an intermediate in Smith's synthesis of normethyljatrophone (**59**)[50] and its lactone analogs.[49]

(1) LDA–PhSeBr
(2) LDA, HCHO
(3) O_3

57 58 59

(6.16)

Treatment of *a*-alkyl-*a*-phenylseleno ketones with 0.5 equivalent of LDA in HMPA at −78° C→25° C results in the intermolecular migration of the PhSe group to the other *a*-carbon. The reaction was studied by Liotta,[51] who also used it in a very efficient synthesis of *cis*-jasmone (**61**, Eq. 6.17).[52]

(1) $LiCuMe_2$
(2) NH_4Cl
(95%)

(1) LDA, HMPA
(2) Br

(100%) 0.5 eq. LDA, HMPA, −78°→ 25° C

H_2O_2
(100%)

(1) MeONa
(2) H_2, Lindlar
(83%)

60 61

(6.17)

Anions resulting from conjugate addition to activated vinylic selenides such as **60**, can also be readily alkylated.[53] Oxidative elimination of the PhSe group of the product regenerates the olefinic bond. Liotta has used this approach in the synthesis of dehydrojasmone (**62**, Eq. 6.18).[5]

SePh (1) $LiCuMe_2$ (2) I (HMPA) (85%) **60** SePh (93%) H_2O_2 **62** + HCl, *n*-BuOH, Δ

(6.18)

Conversion of α-phenylseleno ketones, such as **63**, to their silyl enol ethers (**64**), followed by selective selenium–lithium exchange with metallic lithium and dimethylaminonaphthalene (DMAN) followed by quenching with water, allows the preparation of α-silyl ketones (**65**, Eq. 6.19).[54]

SePh LDA, HMPA, t-$BuMe_2SiCl$ **63** OSi^tBuMe_2 SePh **64** (1) Li, DMAN (2) H_2O (85%) Si^tBuMe_2 **65**

(6.19)

α-Phenylselenopropanoic acid dianion (**66**) has been used in an approach to *trans*-fused α-methylene lactones (**67**), *cis*-fused α-methylene lactones (**68**) and butenolides (**69**), as shown in Eq. 6.20.[55]

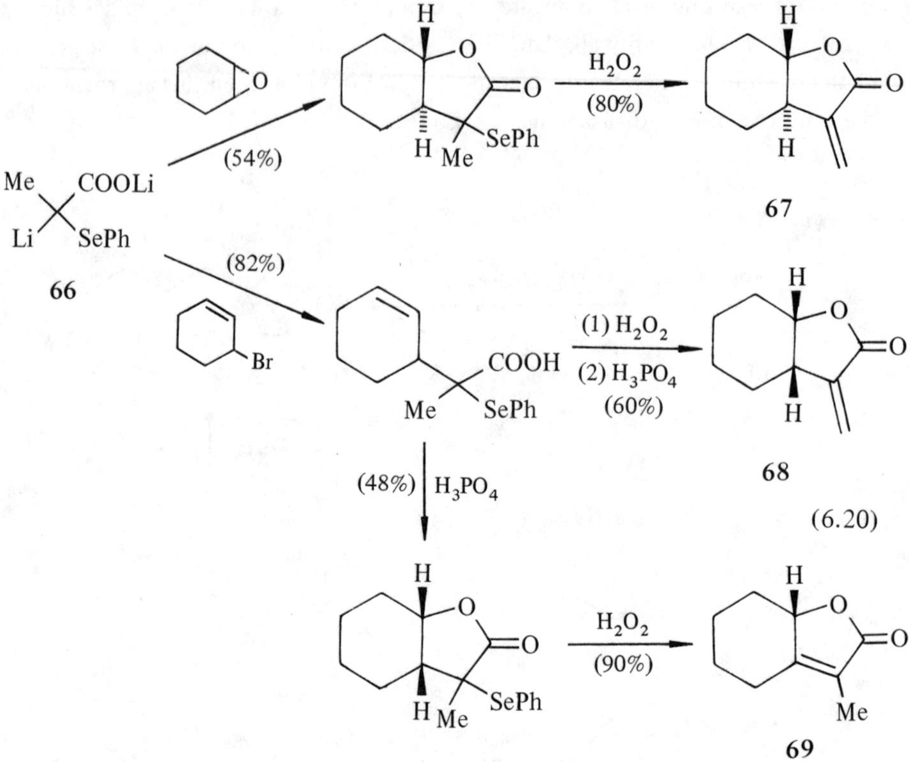

$$(6.20)$$

The reaction of the carbanion derived from penicillin selenoketal derivative **70** with acetaldehyde, to give selectively adduct **71**, was best achieved with methylmagnesium bromide as the base. When *n*-butyllithium was used, a 2:3 mixture of **71** and its stereoisomer was obtained. Reductive removal of the PhSe group gave **72**, which was further elaborated to isopenam derivative **73** (Eq. 6.21).[56]

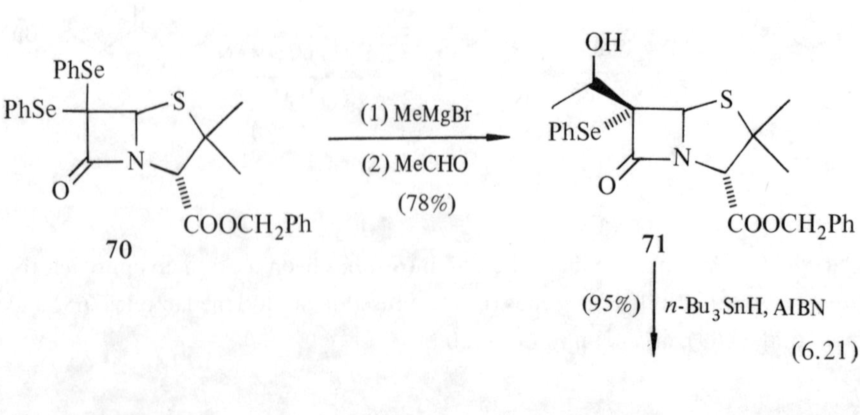

$$(6.21)$$

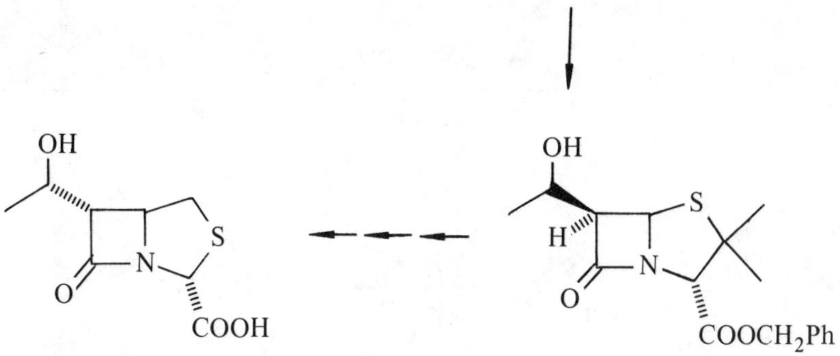

OH

S

N

O

COOH

73

OH

S

H

N

O

$COOCH_2Ph$

72

6.3 Reactions of the carbanions of selenides additionally stabilized by miscellaneous α-substituents

Selenides having carbanion stabilizing groups in the α-position (**74**) can be easily deprotonated with amide bases, such as LDA. Reaction of these additionally stabilized carbanions (**75**) with electrophiles ($E^{\oplus}$) gives highly functionalized selenides (**76**, Eq. 6.22). Oxidative (Ch. 4) or reductive (Ch. 5) removal of the PhSe group from **76**, leads to various useful products. As shown below, several types of such selenides have been investigated.

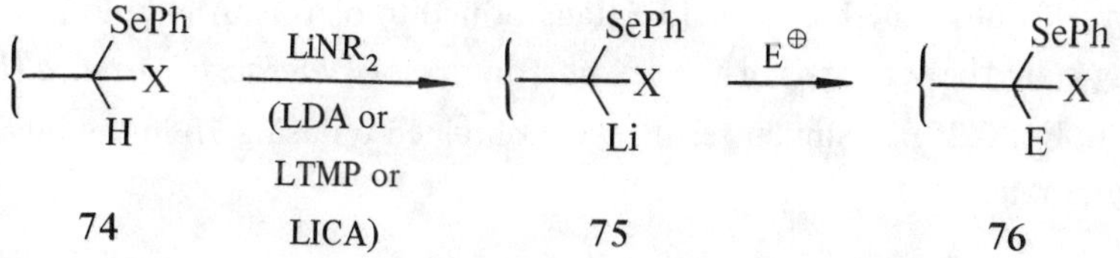

SePh, X, H — $LiNR_2$ (LDA or LTMP or LICA) → SePh, X, Li — $E^{\oplus}$ → SePh, X, E

74 75 76

X = C=C, C≡C, Ar,
SePh, SPh, $SiMe_3$,
CN, COOR, OR (6.22)

The allyl selenide carbanion (**78**), obtained by deprotonation of allyl selenide (**77**)[57, 58], was used by Reich in a method for the synthesis of allylic alcohols (e. g. **79**, Eq. 6.23)[57]. Under similar conditions the chloro-substituted derivative **80**, produces the enones (e. g. **81**, Eq. 6.24)[57]. In some cases, and depending on the electrophile, these allyl carbanions give also some γ-alkylation products[57].

PhSe (77) —LDA→ PhSe–CH(Li)–CH=CH₂ (78) —Ph⌒⌒Br→ (6.23)

Ph⌒⌒⌒CH(SePh)–CH=CH₂ —H_2O_2 (68%)→ Ph⌒⌒⌒CH=CH–CH₂OH (79)

PhSe–CH₂–CH=C(Cl)–CH₃ (80) —(1) LDA (2) Ph⌒Br→ Ph⌒⌒CH(SePh)–CH=C(Cl)–CH₃ (6.24)

—H_2O_2 (85%)→ Ph⌒⌒CH=CH–C(=O)–CH₃ (81)

In the presence of triethylaluminum, carbanion 78 adds to aldehydes, forming β-hydroxyselenides, which can then be transformed to olefins. An application of this sequence in the synthesis of the sex pheromone of *Diparopsis castanea* (82) is shown in Eq. 6.25.[60] Similar results were obtained by using 78 in the presence of trialkylboranes.[94]

(6.25)

PhSe–CH(Li)–CH=CH₂ (78) —Et_3Al→ PhSe–CH=CH–CH₂–$\overset{\ominus}{Al}Et_3$ $Li^{\oplus}$

↓ O=CH–(CH₂)₈–OAc

CH₂=CH–CH(SePh)–CH(OH)–(CH₂)₈–OAc —$H^{\oplus}$ (77%)→

CH₂=CH–CH=CH–(CH₂)₈–OAc 82

The anion of 1,3-bis(phenylseleno)propene (**83**) has been employed by Reich for the synthesis of α,β-unsaturated aldehydes[61] and silyl enones.[61,62] Examples are given in Eq. 6.26[61] and 6.27.[61,62] Interestingly, compound **83** was also found to be a good precursor to selenium-stabilized carbocations.[95]

PhSe SePh (83) —(1) LDA; (2) propylene oxide; (3) Ac_2O→ AcO, SePh, SePh product —H_2O_2 (71%)→ AcO, H, O product (6.26)

PhSe SePh (83) —(1) LDA; (2) Me_3SiCl→ PhSe, SePh, $SiMe_3$ product —H_2O_2 (75%)→ Me_3Si, O, H product; —(1) LDA; (2) isopropyl bromide→ SePh, SePh, $SiMe_3$ product —H_2O_2 (75%)→ O, $SiMe_3$ product (6.27)

α-Phenylselenoenones are obtained with the anion of propargyl selenide (**84**) through the pathway illustrated in Eq. 6.28,[63,64] for the synthesis of **85**, which was converted to the potentially useful diene, as shown.[65]

PhSe (84) —(1) t-BuNHLi; (2) Me_3SiCl→ PhSe, $SiMe_3$ product —mCPBA (56%)→

(6.28)

85

86

Using the dianion **87**, derived from **84** with 2 equivalents of LDA, more highly functionalized enones can be synthesized.[63, 64] Reich has exemplified this methodology in a synthesis of 7-hydroxymyoporone (**88**), presented in Eq. 6.29.[64, 66]

PhSe

2 eq. LDA

PhSe

Li

Li

(1)

(2)

84

87

OH

PhSe

(1) mCPBA

(78%)

(2) Pyridine

X = H, SePh

(1) $LiCuMe_2$
(58%) (2) PhSH, Et_3N
(3) AcOH

88

(6.29)

The carbanion of allenyl phenyl selenide (**89**) was also prepared by Reich, as shown in Eq. 6.30,[64] and found to yield acetylenic alcohols upon alkylation and oxidative removal of the PhSe group. This carbanion, however, gives also the acetylenic products when it is reacted with carbonyl compounds.[64]

PhSe LDA PhSe Li **89** Ph I PhSe

2 LDA Cl PhSe

PhSe— ≡ —

LDA

PhSe— ≡ — Li

H_2O_2

(6.30)

OH Ph

O PhSe Ph

Vinyl selenides (**90**), are now accessible by a large number of methods,[35,67–75] and can be deprotonated with alkyl amide bases (LDA,[18,58,74] KDA,[12] LTMP[58,74]) to give vinyl carbanions (**91**), which are also obtained from ketene selenoacetals (**92**) and *n*-butyllithium.[76,77] These carbanions (**91**) can also be reacted with several kinds of electrophiles (Me_3SiCl,[74] PhSeSePh,[74] RSSR,[74] RX,[12,74,77] Me_2NCHO,[77] CO_2,[74,77] ClCOOR,[77] RCOR,[12,74,77] RCHO[12,77] and epoxides[12]) to give the corresponding vinyl adducts (**93**, Eq. 6.31), which can be converted to various types of selenium-free products (carbonyl compounds,[12,68,70,77,78] olefins,[77] vinyl bromides[77] and allenes[77]).

SeR, H (**90**) —LDA (or KDA or LTMP)→ SeR, Li (**91**) ←*n*-BuLi— SeR, SeR (**92**)

91 —$E^{\oplus}$→ SeR, E (**93**)

(R = Ph, Me, C_6H_4–mCF_3)

(6.31)

Selenide carbanions additionally stabilized by another selenide group, are obtained either by deprotonation of selenoacetals (**4**)[3,12–15] or by Se–Li exchange of selenoorthoesters (**5**, Eq. 6.1).[3,14,15] Alkylation of these carbanions to give selenoketals (**2**), followed by hydrolysis (CuCl, CuO, H_2O, Me_2CO,[12,13] $HgCl_2$, $CaCO_3$, H_2O, MeCN,[13] H_2O_2, THF[13] or PhSe(O)OSe(O)Ph[13]) or reduction (Ph_3SnH,[79] §5.1), leads to ketones and hydrocarbons, respectively.

α-Silyl selenides (**95**) are quite versatile synthetic intermediates. According to Reich, they can be prepared from selenides (**94**) by silylation, and they can be used for the preparation of vinyl silanes (**96**) or silyl ketones (**98**) via the disilyl selenides (**97**) as illustrated in Eq. 6.32.[80]

$$\text{PhCH}_2\text{SePh}\ (\mathbf{94}) \xrightarrow[(2)\ \text{Me}_3\text{SiCl}\ (83\%)]{(1)\ \text{LDA}} \text{Me}_3\text{Si–CH(SePh)Ph}\ (\mathbf{95}) \xrightarrow[(2)\ \text{MeI}\ (77\%)]{(1)\ \text{LDA}} \text{Me}_3\text{Si(Me)C(SePh)Ph}$$

$$\text{Me}_3\text{Si(Me)C(SePh)Ph} \xrightarrow{\text{mCPBA}\ (64\%)} \text{Me}_3\text{SiC(Ph)=CH}_2\ (\mathbf{96})$$

$$\mathbf{95} \xrightarrow[(2)\ \text{Me}_3\text{SiCl}]{(1)\ \text{LDA}\ (51\%)} (\text{Me}_3\text{Si})_2\text{C(SePh)Ph}\ (\mathbf{97}) \xrightarrow[(46\%)]{\text{H}_2\text{O}_2} \text{PhC(O)SiMe}_3\ (\mathbf{98}) \qquad (6.32)$$

Another route to α-silyl selenides, developed by Krief, is shown in Eq. 6.33.[81, 82]

$$\text{HC(SeMe)}_3 \xrightarrow[(2)\ \text{RX}]{(1)\ \text{LDA}} \text{RC(SeMe)}_3 \xrightarrow[(2)\ \text{Me}_3\text{SiCl}]{(1)\ n\text{-BuLi}} \text{Me}_3\text{Si(R)C(SeMe)}_2$$

$$\xrightarrow[(2)\ \text{R}^1\text{X}]{(1)\ n\text{-BuLi}} \text{Me}_3\text{Si(R)C(SeMe)R}^1 \qquad (6.33)$$

Silyl selenide **99** has been developed into a reagent for aldehyde synthesis, as described in Eq. 6.34[83].

$$\text{PhSeCH}_2\text{SiMe}_3\ (\mathbf{99}) \xrightarrow[(2)\ \text{CH}_3(\text{CH}_2)_4\text{Br}]{(1)\ \text{LDA}} \text{CH}_3(\text{CH}_2)_5\text{CH(SePh)SiMe}_3$$

$$\xrightarrow[(80\%)]{\text{H}_2\text{O}_2} \text{CH}_3(\text{CH}_2)_5\text{CHO} \qquad (6.34)$$

Several other α-substituted selenides have been used for the synthesis of substituted olefins. Examples are given in Eqs. 6.35[59, 84], 6.36[59] and 6.37.[85] The utilization of α-phenylseleno Wittig-type reagents for the synthesis of vinyl selenides is discussed in §8.

SePh
Ph
(1) LDA
(2) Br
SePh
Ph
O_3
(78%)
Ph

(6.35)

SeC_6H_4-CF_3-m
OMe
(1) LTMP
(2) PhCOMe
(70%)
ArSe OMe
Ph Me
OH
$SOCl_2$, NEt_3
(73%)
OMe
Ph Me

(6.36)

CN
SePh
NaOH, H_2O, n-Bu_4NI,
Ph X
(81%)
Ph CN
SePh
H_2O_2
Ph CN

(6.37)

6.4 Reactions of the carbanions of selenoxides

Selenoxides stabilize negative charge at the α-carbon in an even higher extent than selenides. Such carbanions react effectively with alkyl halides or carbonyl compounds, forming olefins or allylic alcohols, respectively, after warming up in the presence of an alkylamine. The thermal instability of selenoxides however (Ch. 4), requires their preparation and deprotonation at low temperature. Except methyl phenylselenoxide (MeSe(O)Ph) and benzyl phenyl selenoxide ($PhCH_2Se(O)Ph$), which are stable, other selenoxides are prepared *in situ* by oxidation of selenides at $-78°$ C with ozone or m-chloroperbenzoic acid, followed by addition of LDA, quenching with the electrophile and warming up[84, 86]. Two examples are given in Eqs. 6.38[84, 86] and 6.39[84, 86].

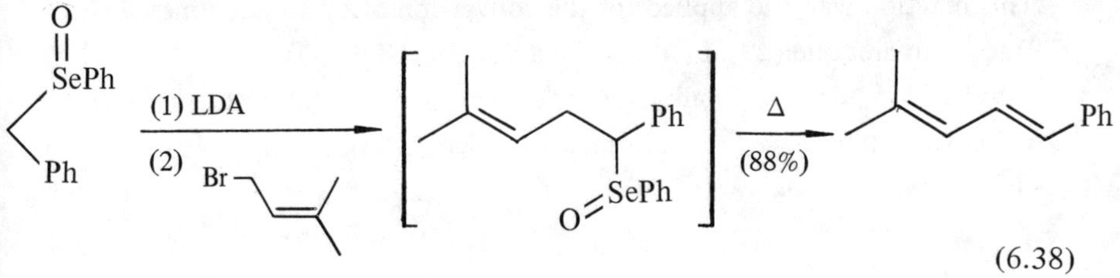

(6.38)

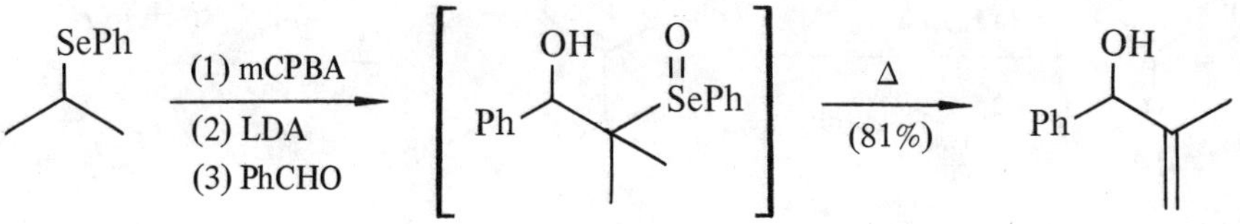

(6.39)

A ring expansion of cyclobutanones, based on the anions of selenoxides was recently reported and is presented in Eq. 6.40[89].

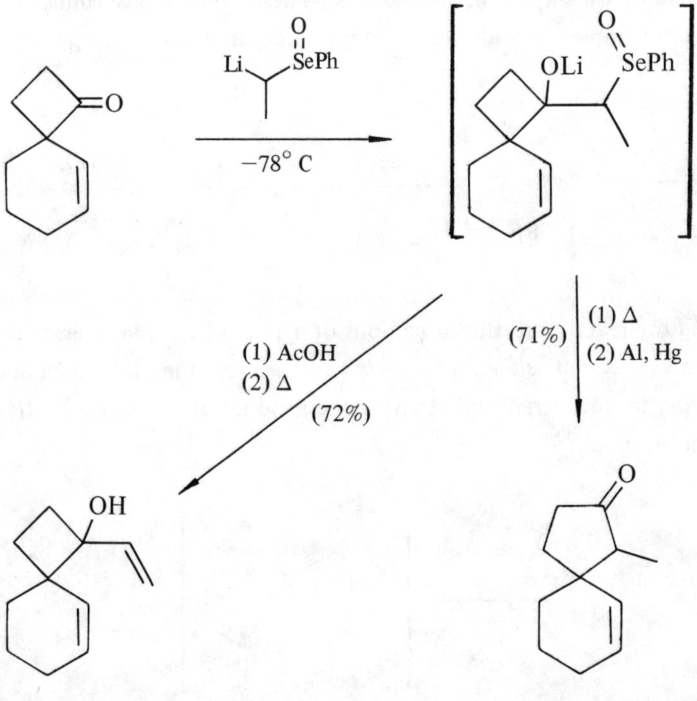

(6.40)

This reaction was also applied for the conversion of cyclobutanone **27** (Eq. 6.9) to α-cuparenone (**29**), by using the anion $Me_2CLiSe(O)Ph$.[89]

Vinyl selenoxides can be alkylated and then thermolyzed in the presence of DABCO to yield acetylenes and sometimes allenes.[90] An example is given in Eq. 6.41.[90]

PhSe(O)CH=CH2 → (1) LDA; (2) PhCH2CH2COCH3 (67%) → PhSe(O)C(=CH2)C(OH)(CH3)CH2CH2Ph → DABCO, 85° C (83%) → HC≡C–C(OH)(CH3)CH2CH2Ph

(6.41)

6.5 *Reactions of selenium ylids*

Selenium ylids are obtained by deprotonation of selenonium salts, which are formed readily by alkylation of selenides. These ylids behave similarly to sulfur ylids. For example, they add to carbonyl compounds forming epoxides (Eq. 6.42).[91]

Ph_2Se → MeI, $AgBF_4$ (60%) → $Me{-}Se^{\oplus}Ph_2\ BF_4^{\ominus}$ → t-BuOK → $^{\ominus}CH_2{-}Se^{\oplus}Ph_2$ → PhCHO, $-Ph_2Se$ → phenyloxirane

(6.42)

During the reaction of the carbanions of α-phenylselenoketones with allyl halides, alkylation at selenium forms *in situ* the selenium ylid which undergoes [2,3]-sigmatropic rearrangement giving the product of formal S_N2' attack, as shown in Eq. 6.43.[92]

$PhCOCH^{\ominus}SePh$ → $ICH_2CH{=}C(CH_3)_2$ → [ylid intermediate: $PhCOCH^{\ominus}{-}Se^{\oplus}Ph{-}CH_2CH{=}C(CH_3)_2$] → $PhCOCH(SePh)C(CH_3)_2CH{=}CH_2$

(6.43)

Such [2,3]-sigmatropic rearrangements seem to proceed rather easily with selenium ylids, as indicated also in Eq. 6.44[36] and 6.45.[93]

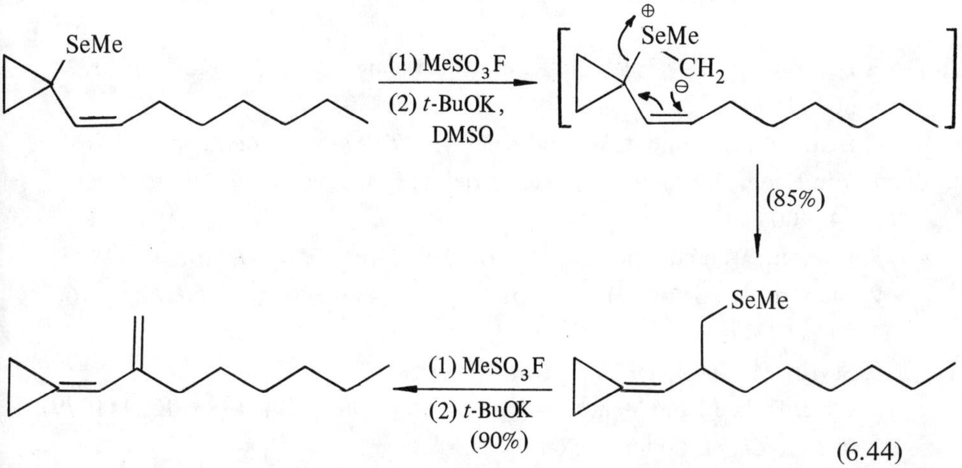

(1) $MeSO_3F$
(2) t-BuOK, DMSO
(85%)
(1) $MeSO_3F$
(2) t-BuOK
(90%)

(6.44)

$MeSO_3F$
(88%)
$NaNH_2$
(43%)

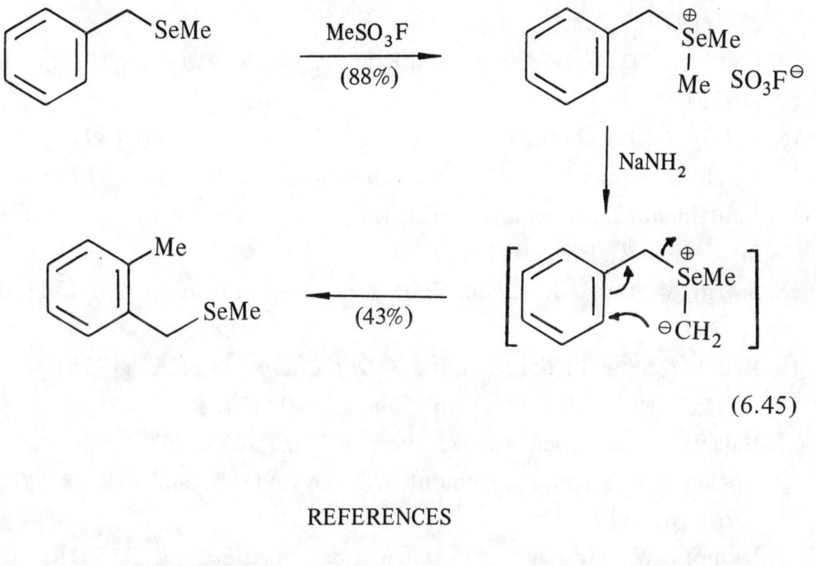

(6.45)

REFERENCES

1. Krief, A., *Tetrahedron,* **36**, 2531 (1980).
2. Reich, H. J., "Organoselenium Oxidations," in Trahanovsky, W. S., (ed.), *Oxidation in Organic Chemistry, Part C*, Academic Press, New York (1978), Chap. 1, p. 1; *Acc. Chem. Res.*, **12**, 22 (1979).
3. Seebach, D., and Peleties, N., *Angew. Chem. Int. Ed. Engl.*, **8**, 450 (1969);

Chem. Ber., **105**, 511 (1972).
4. (a) Seebach, D., and Beck, A. K., *Angew. Chem. Int. Ed. Engl.*, **13**, 806 (1974); (b) Seebach, D., Meyer, N., and Beck, A. K., *Liebigs Ann. Chem.*, 846 (1977).
5. Dumont, W., Bayet, P., and Krief, A., *Angew. Chem. Int. Ed. Engl.*, **13**, 804 (1974).
6. Denis, J. N., Dumont, W., and Krief, A., *Tetrahedron Lett.*, 453 (1976).
7. Van Ende, D., Dumont, W., and Krief, A., *Angew. Chem. Int. Ed. Engl.*, **14**, 700 (1975).
8. Dumont, W., and Krief, A., *Angew. Chem. Int. Ed. Engl.*, **16**, 540 (1977).
9. Dumont, W., Sevrin, M., and Krief, A., *Angew. Chem. Int. Ed. Engl.*, **16**, 541 (1977).
10. Sevrin, M., and Krief, A., *Tetrahedron Lett.*, 187 (1978).
11. Clive, D. L. J., and Menchen, S. M., *J. Org. Chem.*, **44**, 1883, 4279 (1979); *J. C. S. Chem. Comm.*, 356 (1979).
12. Raucher, S., and Koolpe, G. A., *J. Org. Chem.*, **43**, 3794 (1978).
13. Burton, A., Hevesi, L., Dumont, W., Cravador, A., and Krief, A., *Synthesis*, 877 (1979).
14. Van Ende, D., Cravador, A., and Krief, A., *J. Organometall. Chem.*, **177**, 1 (1979).
15. Halazy, S., Lucchetti, J., and Krief, A., *Tetrahedron Lett.*, 3971 (1978).
16. Reich, H. J., and Shah, S. K., *J. Am. Chem. Soc.*, **97**, 3250 (1975).
17. Kauffmann, T., Ahlers, H., Tilhard, H.-J., and Woltermann, A., *Angew. Chem. Int. Ed. Engl.*, **16**, 710 (1977).
18. Sevrin, M., Denis, J. N., and Krief, A., *Angew. Chem. Int. Ed. Engl.*, **17**, 526 (1978).
19. Raucher, S., and Koolpe, G. A., *J. Org. Chem.*, **43**, 4252 (1978).
20. Mitchell, R. H., *J. C. S. Chem. Comm.*, 990 (1975).
21. Halazy, S., and Krief, A., *Tetrahedron Lett.*, 4233 (1979).
22. Anciaux, A., Eman, A., Dumont, W., Van Ende, D., and Krief, A., *Tetrahedron Lett.*, 1613 (1979).
23. Dumont, W., and Krief, A., *Angew. Chem. Int. Ed. Engl.*, **15**, 161 (1976).
24. Rémion, J., and Krief, A., *Tetrahedron Lett.*, 3743 (1976).
25. Rémion, J., and Krief, A., *Tetrahedron Lett.*, 3743 (1976).
26. Halazy, S., Lucchetti, J., and Krief, A., *Tetrahedron Lett.*, 3971 (1978).
27. Halazy, S., and Krief, A., *J. C. S. Chem. Comm.*, 1136 (1979); *Tetrahedron Lett.*, **22**, 1829, 1833 (1981).

28. Labar, D., and Krief, A., *J. C. S. Chem. Comm.,* 564 (1982).
29. Halazy, S., and Krief, A., *J. C. S. Chem. Comm.,* 1200 (1982).
30. Labar, D., Laboureur, J. L., and Krief, A., *Tetrahedron Lett.,* **23**, 983 (1982).
31. Van Ende, D., and Krief, A., *Tetrahedron Lett.,* 457 (1976).
32. Sevrin, M., and Krief, A., *Tetrahedron Lett.,* 585 (1980).
33. Labar, D., Dumont, W., Hevesi, S., and Krief, A., *Tetrahedron Lett.,* 1145 (1978).
34. DiGiamberardino, T., Halazy, S., Dumont, W., and Krief, A., *Tetrahedron Lett.,* **24**, 3413 (1983).
35. Reich, H. J., and Chow, F., *J. C. S. Chem. Comm.,* 790 (1975).
36. Halazy, S., and Krief, A., *Tetrahedron Lett.,* **22**, 2135 (1981).
37. Halazy, S., and Krief, A., *Tetrahedron Lett.,* **22**, 4341 (1981).
38. (a) Trost, B. M., Keeley, D. E., Arndt, H. C., and Bogdanowicz, M. J., *J. Am. Chem. Soc.,* **99**, 3088 (1977); (b) Trost, B. M., Keeley, D. E., Arndt, H. C., Rigby, J. H., and Bogdanowicz, M. J., *J. Am. Chem. Soc.,* **99**, 3080 (1977); (c) Trost, B. M., Keeley, D. E., and Bogdanowicz, M. J., *J. Am. Chem. Soc.,* **95**, 3068 (1972).
39. Halazy, S., Zutterman, F., and Krief, A., *Tetrahedron Lett.,* **23**, 4385(1982).
40. Schauder, J. R., and Krief, A., *Tetrahedron Lett.,* **23**, 4389 (1982).
41. Baudat, R., and Petrzilka, M., *Helv. Chim. Acta,* **62**, 1406 (1979).
42. Corey, E. J., and Kim, C. U., *J. Am. Chem. Soc.,* **94**, 7586 (1972).
43. Lucchetti, J., and Krief, A., *Compt. Rend.,* **288C**, 537 (1979).
44. Posner, G. H., and Chapdelaine, M. J., *Tetrahedron Lett.,* 3227 (1977).
45. Barton, D. H. R., Lester, D. J., Moterwell, W., and Barros Papoula, M. T., *J. C. S. Chem. Comm.,* 246 (1980).
46. Halazy, S., and Krief, A., *Tetrahedron Lett.,* **21**, 1997 (1980).
47. (a) Grieco, P. A., Nishizawa, M., Oguri, T., Burke, S. D., and Marinovic, N., *J. Am. Chem. Soc.,* **99**, 5773 (1977); (b) Grieco, P. A., Nishizawa, M., Burke, S. D., and Marinovic, N., *J. Am. Chem. Soc.,* **98**, 1612 (1976).
48. Nishizawa, M., Grieco, P. A., Burke, S. D., and Metz, W., *J. C. S. Chem. Comm.,* 76 (1978).
49. Smith, A. B., III, and Malamas, M. S., *J. Org. Chem.,* **47**, 3442 (1982).
50. Smith, A. B., III,Guaciaro, M. A., Schow, S. R., Wovkulich, P. M., Toder, B. H., and Hall, T. W., *J. Am. Chem. Soc.,* **103**, 219 (1981).
51. Liotta, D., Saindane, M., and Brothers, D., *J. Org. Chem.,* **47**, 1598 (1982).
52. Liotta, D., Barnum, C. S., and Saindane, M., *J. Org. Chem.,* **46**, 4301 (1981).
53. Zima, G., Barnum, C., and Liotta, D., *J. Org. Chem.,* **45**, 2736 (1980).

54. Kuwajima, I., and Takeda, R., *Tetrahedron Lett.,* **22**, 2381 (1981).
55. Petragnani, N., and Ferraz, H. M. C., *Synthesis*, 476 (1978).
56. Hirai, K., Iwano, Y., and Fujimoto, K., *Tetrahedron Lett.,* **23**, 4021, 4027 (1982); *Heterocycles,* **17**, 201 (1982).
57. Reich, H. J., *J. Org. Chem.,* **40**, 2570 (1975).
58. Reich, H. J., and Willis, W. W., Jr., *J. Org. Chem.,* **45**, 5227 (1980).
59. Reich, H. J., Chow, F., and Shah, S. K., *J. Am. Chem. Soc.,* **101**, 6638 (1979).
60. Yamamoto, Y., Saito, Y., and Maruyama, K., *Tetrahedron Lett.,* **23**, 4597 (1982).
61. Reich, H. J., Clark, M. C., and Willis, W. W., Jr., *J. Org. Chem.,* **47**, 1619 (1982).
62. Reich, H. J., Olson, R. E., and Clark, M. C., *J. Am. Chem. Soc.,* **102**, 1423 (1980).
63. Reich, H. J., and Shah, S. K., *J. Am. Chem. Soc.,* **99**, 263 (1977).
64. Reich, H. J., Shah, S. K., Gold, P. M., and Olson, R. E., *J. Am. Chem. Soc.,* **103**, 3112 (1981).
65. Reich, H. J., Rusek, J. J., and Olson, R. E., *J. Am. Chem. Soc.,* **101**, 2225 (1979).
66. Reich, H. J., Gold, P. M., and Chow, F., *Tetrahedron Lett.,* 4433 (1979).
67. Petrargani, N., Rodrigues, R., and Comassetto, J. V., *J. Organometall. Chem.,* **114**, 281 (1976).
68. Sevrin, M., Dumont, W., and Krief, A., *Tetrahedron Lett.,* 3835 (1977).
69. Raucher, S., *J. Org. Chem.,* **42**, 2950 (1977).
70. Dumont, W., Sevrin, M., and Krief, A., *Tetrahedron Lett.,* 183 (1978).
71. Raucher, S., Hansen, M. R., and Colter, M. A., *J. Org. Chem.,* **43**, 4885 (1978).
72. Dumont, W., Van Ende, D., and Krief, A., *Tetrahedron Lett.,* 485 (1979).
73. Feiring, A. E., *J. Org. Chem.,* **45**, 1958 (1980).
74. Reich, H. J., Willis, W. W., Jr., and Clark, P. D., *J. Org. Chem.,* **46**, 2775 (1981).
75. Denis, J. N., and Krief, A., *Tetrahedron Lett.,* **23**, 3407 (1982).
76. Grobel, B.-T., and Seebach, D., *Chem. Ber.,* **110**, 852, 867 (1977).
77. Denis, J. N., and Krief, A., *Tetrahedron Lett.,* **23**, 3411 (1982).
78. Hevesi, L., Piquard, J.-L., and Wautier, H., *J. Am. Chem. Soc.,* **103**, 870(1981).
79. (a) Clive, D. L. J., Chittattu, G. J., Farina, V., Kiel, W. A., Menchen, S. M., Russell, C. G., Singh, A., Wong, C. K., and Curtis, N. J., *J. Am. Chem. Soc.,* **102**, 4438 (1980); (b) Clive, D. L. J., Chittattu, G. J., and Wong, C. K., *J.*

C. S. Chem. Comm., 41 (1978).

80. Reich, H. J., and Shah, S. K., *J. Org. Chem.*, **42**, 1773 (1977).
81. Van Ende, D., Dumont, W., and Krief, A., *J. Organometall. Chem.*, **149**, C10 (1978).
82. Halazy, S., Dumont, W., and Krief, A., *Tetrahedron Lett.*, **22**, 4737 (1981).
83. Sachdev, K., and Sachdev, H. S., *Tetrahedron Lett.*, 4223 (1976).
84. Reich, H. J., and Shah, S. K., *J. Am. Chem. Soc.*, **97**, 3250 (1975).
85. Masuyama, Y., Ueno, Y., and Okawara, M., *Chemistry Lett.*, 835 (1977).
86. Reich, H. J., Shah, S. K., and Chow, F., *J. Am. Chem. Soc.*, **101**, 6648 (1979).
87. Gadwood, R. C., *J. Org. Chem.*, **48**, 2098 (1983); see also Gadwood, R. C., and Lett, R. M., *J. Org. Chem.*, **47**, 2268 (1982).
88. Reich, H. J., and Willis, W. W., Jr., *J. Am. Chem. Soc.*, **102**, 5968 (1980).
89. Gadwood, R. C., *J. Org. Chem.*, **48**, 2098 (1983).
90. Reich, H. J., and Willis, W. W., Jr., *J. Am. Chem. Soc.*, **102**, 5967 (1980).
91. Dumont, W., Bayet, P., and Krief, A., *Angew. Chem. Int. Ed. Engl.*, **13**, 274 (1974).
92. Reich, H. J., and Cohen, M. L., *J. Am. Chem. Soc.*, **101**, 1307 (1979).
93. Gassman, P. G., Miura, T., and Mossman, A., *J. C. S. Chem. Comm.*, 558 (1980).
94. Yamamoto, Y., Saito, Y., and Maruyama, K., *J. Org. Chem.*, **48**, 5408 (1983).
95. (a) Renard, M., and Hevesi, L., *Tetrahedron Lett.*, **24**, 3911 (1983); (b) Halazy, S., and Hevesi, L., *J. Org. Chem.*, **48**, 5243 (1983).

CHAPTER 7

Organoselenium-induced Cyclizations

Addition of electrophilic organoselenium reagents (PhSeX) to olefins (**1**), followed by intramolecular capture of the resulting selenonium species (**2**) by an internal nucleophile (Nu), results in the formation of a new ring and the attachment of the PhSe group to the molecule (**3**, Eq. 7.1).

$$\mathbf{1}\ \xrightarrow{\text{PhSeX}}\ [\mathbf{2}]\ \xrightarrow{-\text{HX}}\ \mathbf{3} \qquad (7.1)$$

NuH = OH, COOH, SH, SAc, NHCOOEt, CH_2SnMe_3, CH=CR_2, etc.

Depending on the nature of the internal nucleophile (Nu), the product of this process (termed *cyclofunctionalization*)[1] can be a cyclic ether, a lactone, an S- or N-heterocycle or a carbocyclic compound. All of these cases are examined in this chapter.

7.1 Synthesis of cyclic ethers

Cyclic ethers are quite common among natural products. The most useful and general synthesis of these compounds involves cyclization of unsaturated alcohols, induced by electrophilic reagents such as halogen and the salts of Hg, Pb, Pd and Tl. The use of electrophilic organoselenium reagents in this transformation was introduced by Nicolaou,[2–5] and independently by Clive.[1,6] The reaction (termed *phenylselenoetherification*)[3] proved to be very useful, due to the mildness and diversity of organoselenium compounds. It consists in treating the unsaturated alcohol (**4**) at low temperature, with phenylselenenyl chloride (PhSeCl),[1–3,6] *N*-phenylselenosuccinimide (N-PSS)[4] or *N*-phenylseleno-phthalimide (N-PSP)[4] to give the phenylselenoether (**5**). Oxidative elimination of the PhSe group proceeds preferably away from the oxygen (§4.2), leading to allylic cyclic ethers (**6**), while reductive cleavage of the PhSe group (§5.1) gives the saturated cyclic ethers (**7**, Eq. 7.2). The formation of **6** is a unique feature of the sequence, complementary to the preference of halogen-based cyclizations to give the corresponding vinyl cyclic ethers.[3]

OH PhSeCl −78°C O SePh [O] O [H] O

4 **5** **6** **7**

(7.2)

The method was used by Corey[7] and by Nicolaou[8] in the synthesis of biologically interesting analogs of prostacyclin such as PGI_1 (**9**, with two C-6 isomers) and iso-PGI_2 (**10**, with two C-6 isomers) from prostaglandin F_{2a} (**8**), as shown in Eq. 7.3.[8]

COOMe OH OH OH PhSeCl (80%) PhSe O COOMe OH OH

8

(7.3)

(1) n-Bu_3SnH, AIBN, PhMe, Δ
(2) LiOH
(70%)

(95%)
(1) H_2O_2
(2) LiOH

COOH
O
OH
OH
9

COOH
O
OH
OH
10

(7.3)

Joullié has also used the phenylselenoetherification reaction for the synthesis of of muscarine analog **11** (Eq. 7.4).[3,9]

O
OH
PhSeCl, Et_3N
(80%)
O
O
SePh
$NaBH_4$
(80%)

HO
O
SePh
(1) Ac_2O, pyr–DMAP
(2) O_3
AcO
O

HO
O
$\overset{\oplus}{N}Me_3$
$I^{\ominus}$
11

(7.4)

The method proved to be very useful in Hoye's synthesis of ancistrofuran (**13**, Eq. 7.5),[10] in which it was used to construct both the tetrahydrofuran and the butenolide moieties of intermediate **12**. When diene **14** was subjected to the reaction, a product of 1,4-addition (**15**) was obtained, as shown in Eq. 7.6.[10]

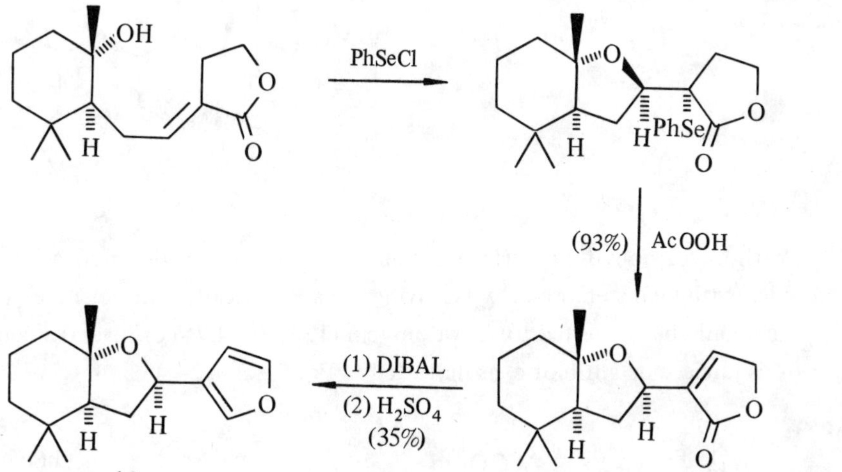
OH
PhSeCl
O
PhSe
(93%)
AcOOH
(1) DIBAL
(2) H_2SO_4
(35%)
H
O

13 **12** (7.5)

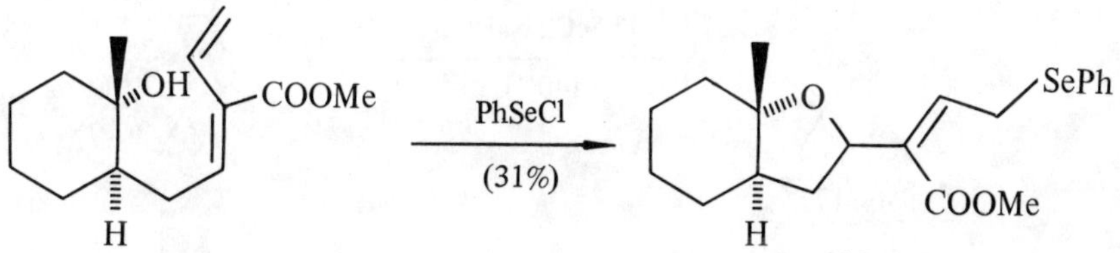
OH
COOMe
PhSeCl
(31%)
SePh
COOMe
H
O

14 **15**

(7.6)

Kraus has also used the method for the synthesis of the tetrahydrofuran ring of an intermediate in a projected synthesis of guasimarin (**16**, Eq. 7.7).[11]

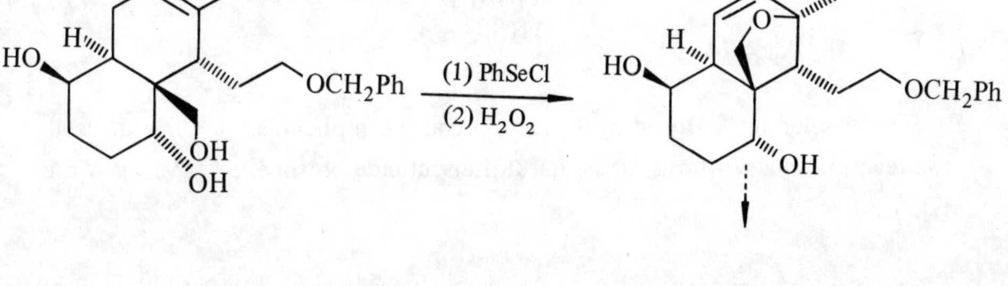
HO
H
OCH_2Ph
OH
OH
(1) PhSeCl
(2) H_2O_2
O

(7.7)

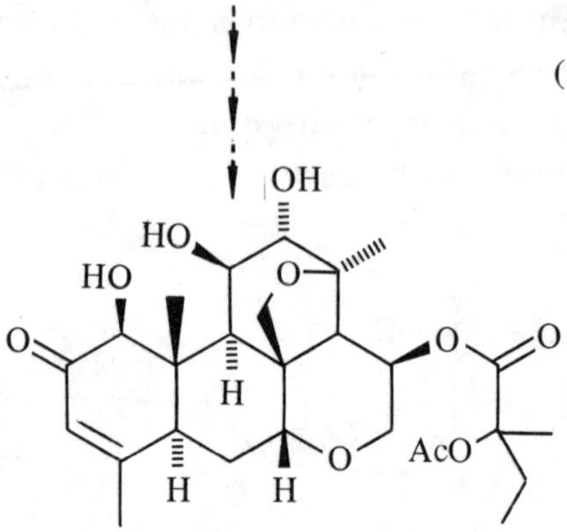

16

With the carbohydrate-derived α,β-unsaturated ester **17**, the cyclization proceeded with high stereoselectivity, giving the anomerically pure selenide (**18**) which could be converted to showdomycin (**135**, Eq. 4.74) or several analogs of this nucleoside antibiotic, as indicated in Eq. 7.8.[12]

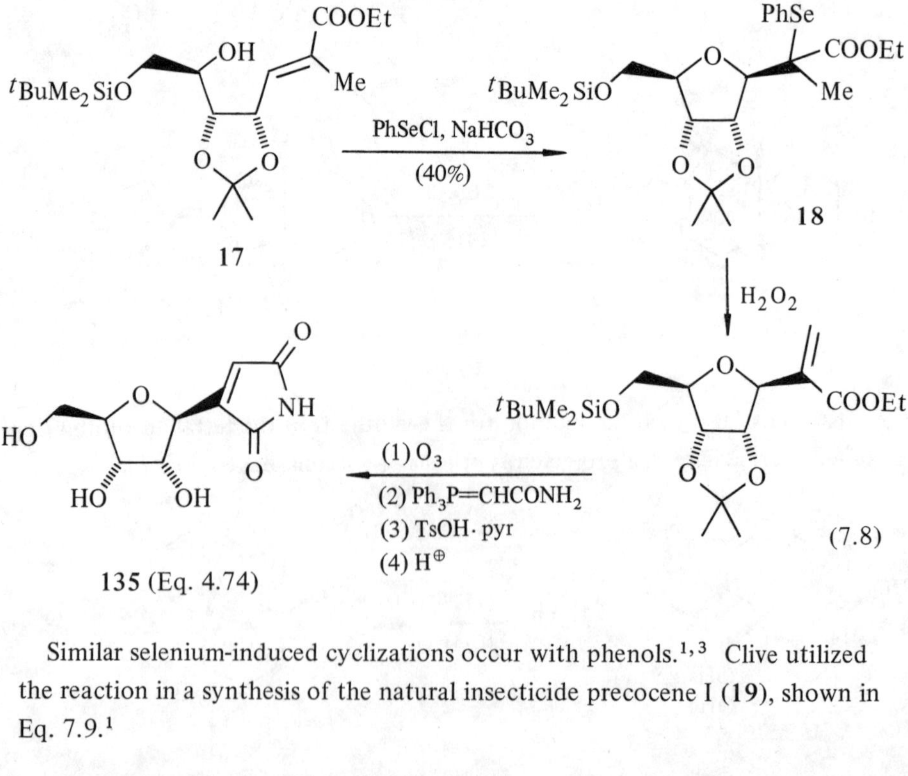

PhSeCl, $NaHCO_3$ (40%)

17 **18**

H_2O_2

(1) O_3
(2) $Ph_3P{=}CHCONH_2$
(3) TsOH· pyr
(4) $H^{\oplus}$

135 (Eq. 4.74)

(7.8)

Similar selenium-induced cyclizations occur with phenols.[1,3] Clive utilized the reaction in a synthesis of the natural insecticide precocene I (**19**), shown in Eq. 7.9.[1]

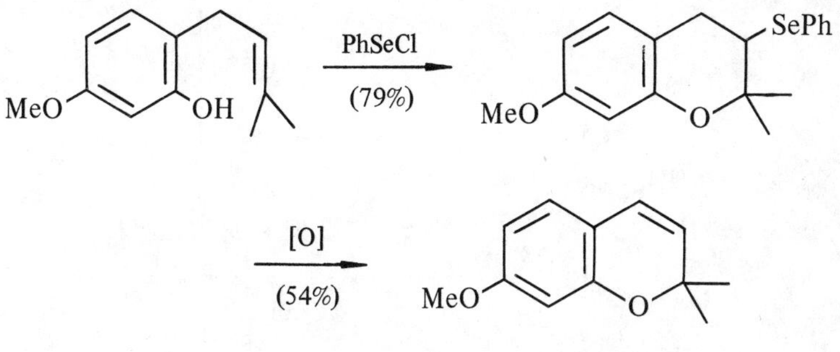

PhSeCl (79%)

MeO OH → MeO O SePh

[O] (54%)

MeO O

19

(7.9)

An example of reductive removal of the PhSe group is found in the synthesis of demethoxy analogs of eleutherin and isoeleutherin (Eq. 7.10).[13]

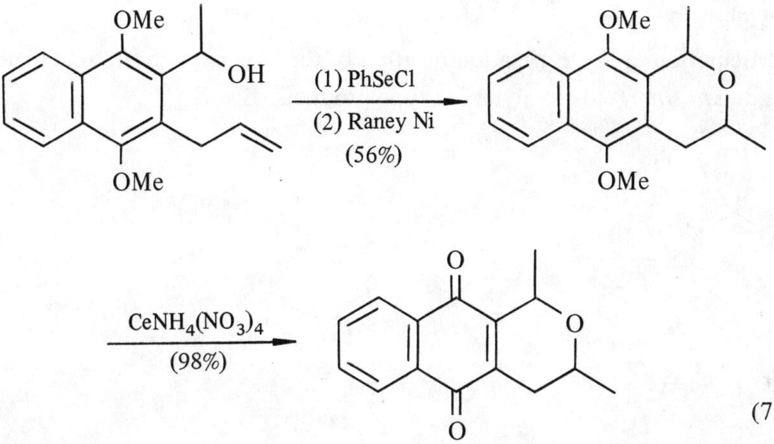

OMe OH OMe → (1) PhSeCl (2) Raney Ni (56%) → OMe O OMe

$CeNH_4(NO_3)_4$ (98%) → O O O

(7.10)

In an approach to the antibiotic bicyclomycin (23, Eq. 7.11),[14] selenide 21, obtained by cyclization of 20, was transformed to the acetoxyselenide (22) via a Pummerer-type rearrangement. This functionality is equivalent to an aldehyde and is readily transformed to it by treatment with sodium hydroxide.[14]

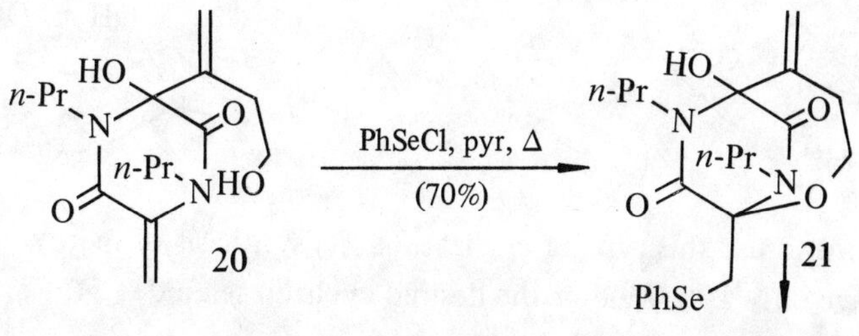

HO n-Pr N O n-Pr N HO O 20

PhSeCl, pyr, Δ (70%)

HO n-Pr N O n-Pr N O O PhSe 21

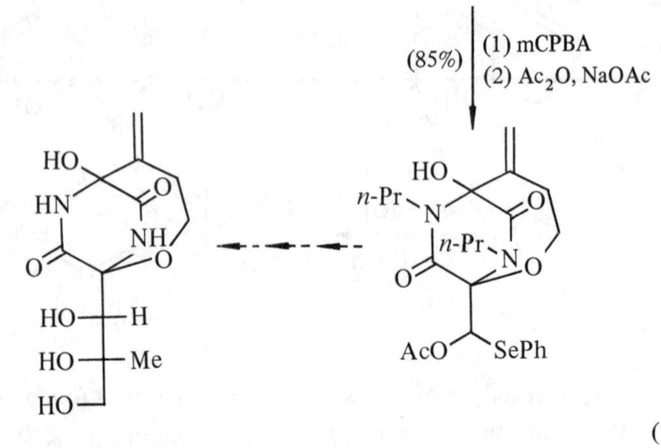

(85%) (1) mCPBA (2) Ac_2O, NaOAc

HO, HN, O, NH, O, O, HO—H, HO—Me, HO

HO, n-Pr—N, O, n-Pr—N, O, O, AcO, SePh

(7.11)

23 22

Participation in the cyclization reaction of the hydroxyl group of a hemiacetal generated *in situ*, leads to glycoside-type products (Eq. 7.12).[15]

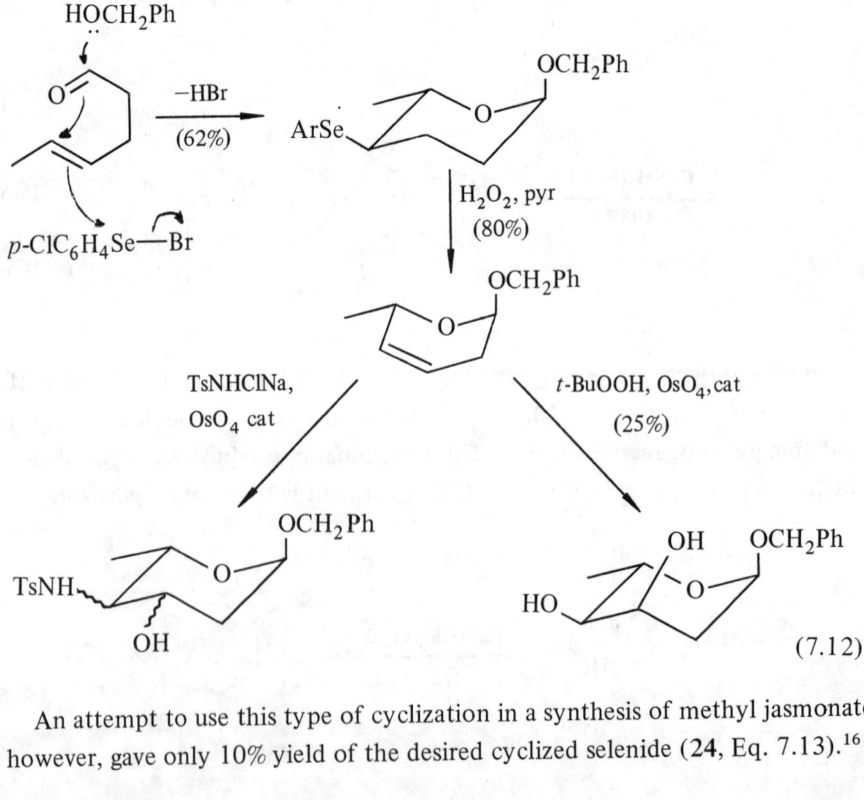

$HOCH_2Ph$

O

$-HBr$ (62%)

$p\text{-}ClC_6H_4Se—Br$

OCH_2Ph, ArSe, O

H_2O_2, pyr (80%)

OCH_2Ph, O

$TsNHClNa$, OsO_4 cat

$t\text{-}BuOOH$, OsO_4, cat (25%)

OCH_2Ph, TsNH, O, OH

OH, OCH_2Ph, HO, O

(7.12)

An attempt to use this type of cyclization in a synthesis of methyl jasmonate, however, gave only 10% yield of the desired cyclized selenide (**24**, Eq. 7.13).[16]

CHO / CHO → PhSeBr, MeOH, Na_2CO_3 (10%) → OMe, O, PhSe, CHO **24** (7.13)

Ley has successfully employed the phenylselenoetherification reaction for the synthesis of spiroketals and 1,3-disubstituted tetrahydropyrans, starting from lactol-type hydroxy compounds (Eq. 7.14)[17] and enolic hydroxy compounds (Eq. 7.15),[18] respectively. The final products of the sequence, *i.e.* **25**, **26** and **27**, are naturally occurring compounds.

OH, O → N-PSP, $ZnBr_2$ cat (78%) → PhSe + SePh

(90%) Raney Ni

25 + **26** (7.14)

O, COOR → N-PSP, $SnCl_4$ cat; R = Me (84%); R = CH_2Ph (57%) → PhSe, O, COOR →

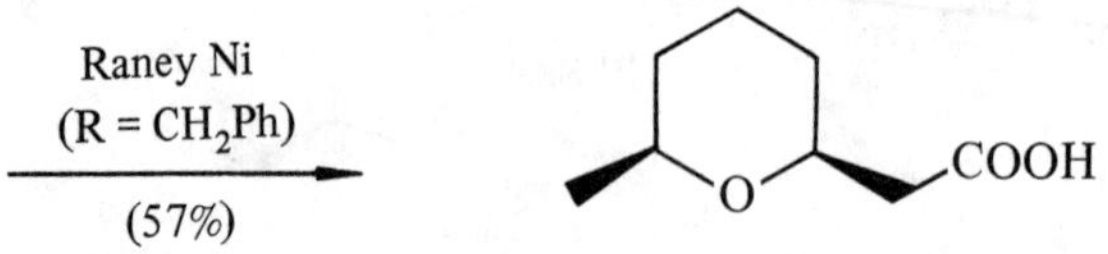

Raney Ni
($R = CH_2Ph$)
(57%)

COOH

27

(7.15)

Petrzilka has used the reaction to prepare cyclic ketene acetal **28**, which undergoes Claisen rearrangement *in situ*, forming the macrolide phoracantholide J (**29**, Eq. 7.16).

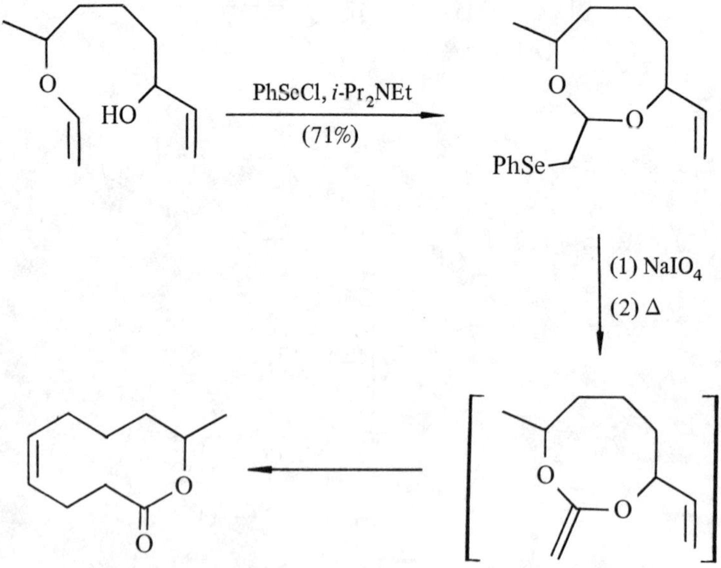

HO

PhSeCl, *i*-Pr_2NEt
(71%)

PhSe

(1) $NaIO_4$
(2) Δ

29

28

(7.16)

The phenylselenoetherification reaction has also been used for purposes of structural determination (Eq. 7.17[20] and Eq. 7.18[21]) and for differentiation of hydroxyl groups (Eq. 7.19).[22]

H
O
O
HO
PhSeCl
SePh
H
O
O
O
n-Bu$_3$SnH,
AIBN, Δ
H
O
O
O
H
O
HO
PhSeCl
H
O
O
PhSe
O
n-Bu$_3$SnH,
AIBN, Δ

(7.17)

COOMe
OH
OH
PhSeCl
(80%)
COOMe
SePh
O
OH

(7.18)

OH
HO
PhSeCl
(85%)
PhSe
OH
O

(7.19)

Reversal of the phenylselenoetherification, *i.e.* reformation of the olefinic alcohols (**4**) from the phenylseleno cyclic ethers (**5**), has been accomplished by Nicolaou[23] and Clive.[24] The reaction (Eq. 7.20) involves the use of sodium in liquid ammonia[23] or trimethylsilyl chloride and sodium iodide in acetonitrile[24] and makes possible the utilization of the selenium cyclization as a means of simultaneous protection of both an olefin and a hydroxyl group. For example, symmetrical diol **30** was nicely functionalized on one side to give alcohol **31** (Eq. 7.21).[25]

O SePh **5** —[$Na–NH_3(l)$ / Me_3SiCl, NaI, MeCN]→ OH **4** (7.20)

OH OH **30** —[PhSeCl (70%)]→ PhSe O OH (7.21)

↓ (1) $SO_3 \cdot$pyr, Et_3N (2) $Ph_3P{=}CH(CH_2)_4Me$

PhSe O —[Me_3SiCl, NaI (70%)]→ OH **31**

Application of selenium-induced cyclizations to α-allenic alcohols (**32**) gives dihydrofurans (**33**, Eq. 7.22).[26]

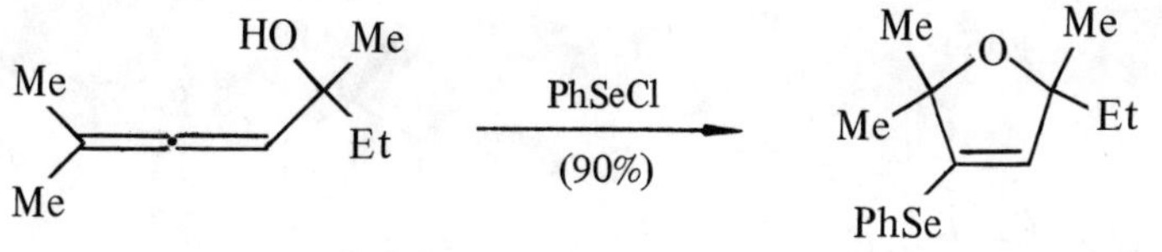

(7.22)

32 33

The corresponding reactions with alkynyl alcohols do not take place.[27] The products obtained result from an irreversible 1,2-addition of the elements of PhSe–Cl to the acetylenic bond (Eq. 7.23).[27]

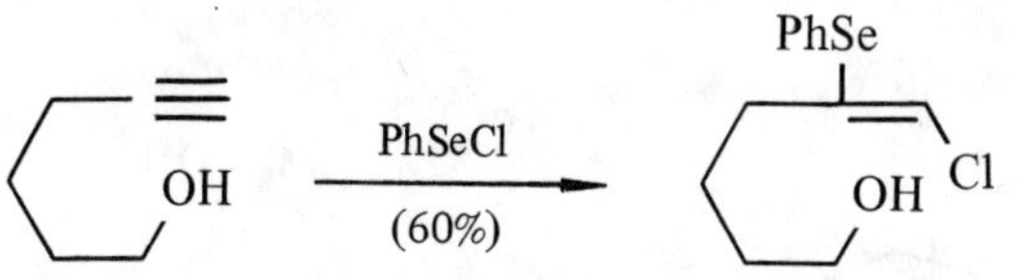

(7.23)

An alternative route to cyclic ethers from acyclic diolefins was also found.[4, 28] It consists in treating the diolefin (**34**) with "PhSeOH" (generated *in situ*) to give cyclic ethers bearing two PhSe groups (**35**, Eq. 7.24). Elimination of the two PhSe groups oxidatively or reductively leads to useful products.[28]

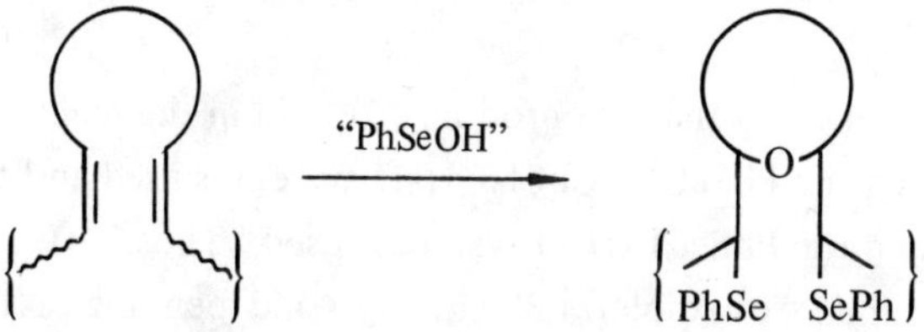

(7.24)

34 **35**

The method was applied in a two-step conversion of limonene (**36**) to eucalyptole (**37**) and *cis*-terpin (**38**), according to Eq. 7.25.[25]

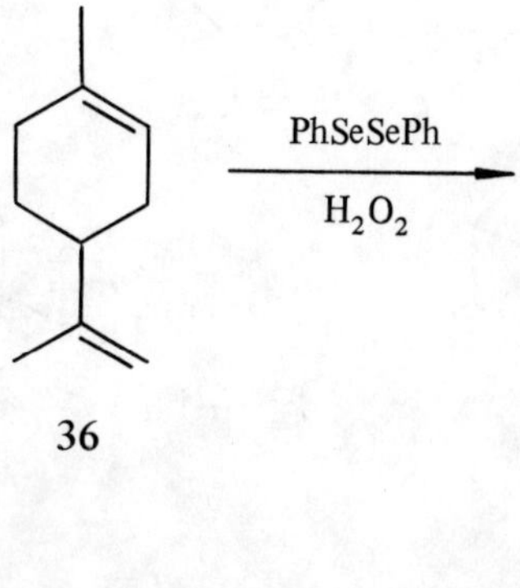

(7.25)

36

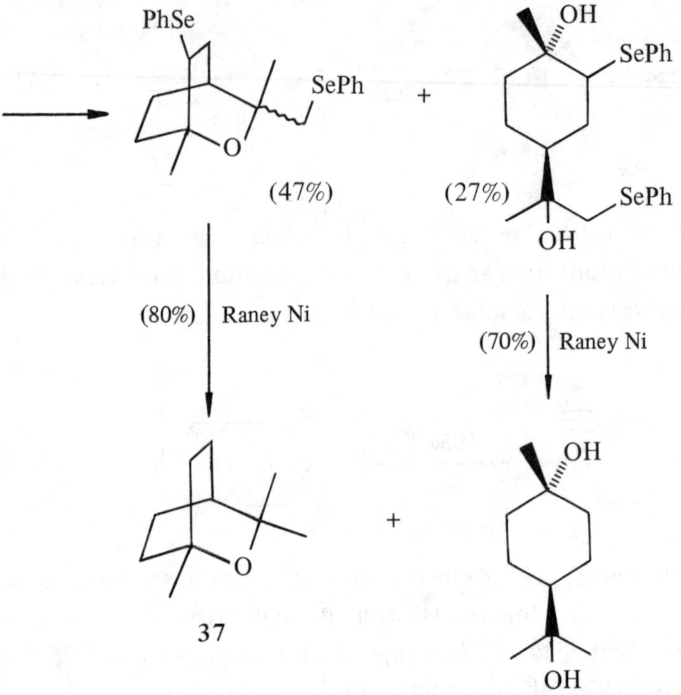

PhSe SePh (47%) + OH SePh (27%) SePh OH

(80%) Raney Ni (70%) Raney Ni

37 + OH OH 38 (7.25)

Uemura has reported that the products formed in the reaction are solvent dependent when the $PhSeCN-CuCl_2-H_2O$ system is used and temperature dependent when the $PhSeCl-H_2O$ system is used. Thus, different ratios of **40**:**41** are obtained from diolefin **39** (Eq. 7.26), depending on the conditions used (Table 7.1).

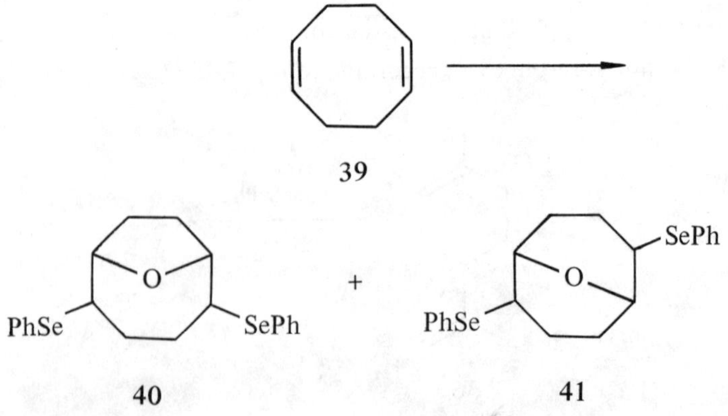

39 (7.26)

PhSe O SePh + PhSe O SePh

40 41

Table 7.1. Ratios of **40**:**41** obtained by oxyselenation of **39**.

Reagent	*Solvent*	*Temperature (°C)*	*Yield (percent)*	**40**:**41** *ratio*	*Ref.*
N-PSS, H_2O	CH_2Cl_2	25	97	100:0	4
N-PSP, H_2O	CH_2Cl_2	25	95	100:0	4
PhSe SePh, H_2O_2	CH_2Cl_2	25	90	100:0	28
PhSeCN, $CuCl_2$	MeOH	64	62	95:5	29
PhSeCN, $CuCl_2$	*t*-BuOH	82	44	38:62	29
PhSeCN, $CuCl_2$	THF–H_2O	60	68	0:100	29
PhSeCl, H_2O	MeCN	25	92	88:12	30
PhSeCl, H_2O	MeCN	76	94	3:97	30

The combination of phenylselenoetherification of a diene with the reverse reaction at one site, allowed the conversion of geranyl acetate (**42**) to linalool (**43**) in 45% overall yield, as shown in Eq. 7.27.[28]

(7.27)

A different type of cycloetherification, based on a selenium-mediated epoxidation, was reported by Kametani.[31] This approach was used in the biomimetic conversion of geraniol (**44**) to *trans*- and *cis*-linalyl oxides (**47** and **48**, Eq. 7.28).[31] Transformation of geraniol (**44**) to *o*-nitrophenyl selenide (**45**), followed by oxidation with hydrogen peroxide, gave the corresponding selenoxide, which underwent facile [2,3]-sigmatropic rearrangement (§4.5), forming a tertiary allylic alcohol and liberating *o*-nitrobenzeneselenenic acid (o–$NO_2C_6H_4SeOH$). Further oxidation of the selenium with hydrogen peroxide formed the seleninic acid, o-$NO_2C_6H_4Se(O)OH$, which mediated the epoxidation of the trisubstituted olefin, giving the epoxy allylic alcohol (**46**). This product could be isolated if the oxidation was done in the presence of pyridine. Acid-catalyzed cyclization of the epoxy alcohol, however, led to a mixture of the *trans* and *cis* linalyl oxides (**47** and **48**). The epoxidation of olefins with ArSe(O)OH seems to be a general and efficient process , as shown in the conversion of farnesol (**49**) to diepoxynerolidol (**50**, Eq. 7.28) [31] and as discussed in §2.5.

OH

o-$NO_2C_6H_4SeCN$

n-Bu_3P

(89%)

NO_2

Se

44

45

H_2O_2 (78%)

H_2O_2, pyridine (96%)

HO O + HO O

TsOH

(87%)

OH

O

47

48

46

OH

mCPBA, $NaHCO_3$

(97%)

(7.28)

(7.29)

Sinay has recently reported a synthesis of carbohydrate spiro-orthoesters (**53**) based on the intramolecular reaction of a hydroxyl group with a ketene-acetal, formed *in situ* by selenoxide elimination (Eq. 7.30).[32] The required selenide (**52**) was prepared from glycol **51** by a "glycosyloxyselenation" reaction. This type of spiro-orthoester functionality occurs in the orthosomycin family of glycoside antibiotics.

(7.30)

7.2 Synthesis of lactones

The synthesis of lactones has attracted considerable attention due to their widespread occurrence in nature and to their great synthetic utility. The most common approach to lactones involves the cyclization of open-chain hydroxy carboxylic acids. The method has been used widely in the synthesis of macrolides.[33] Alternatively, lactones have been synthesized by lactonization of unsaturated carboxylic acids initiated by electrophilic addition of various electrophiles, such as halogens, to the olefinic bond.The use of electrophilic organoselenium species for the induction of lactonization, termed *phenylselenolactonization*,[34] was developed by Nicolaou[4, 5, 34, 35] and studied independently by Clive.[36, 37] The method consists in treating the unsaturated acids (**54**) with PhSeCl, PhSeBr, N-PSS or N-PSP at low temperature, with or without a base catalyst such as triethylamine, pyridine or anhydrous potassium carbonate (Eq. 7.31).[34, 35]

The reaction proceeds with *trans*-addition of the carboxylic acid and the PhSe group to the olefin, giving the phenylselenolactone (**55**, Eq. 7.31). Further elaboration of the selenide function oxidatively, preferably with elimination of the selenoxide away from the oxygen (§4.2), or reductively (§5.1), gives selenium-free unsaturated lactones (**56**) or saturated lactones (**57**), respectively.

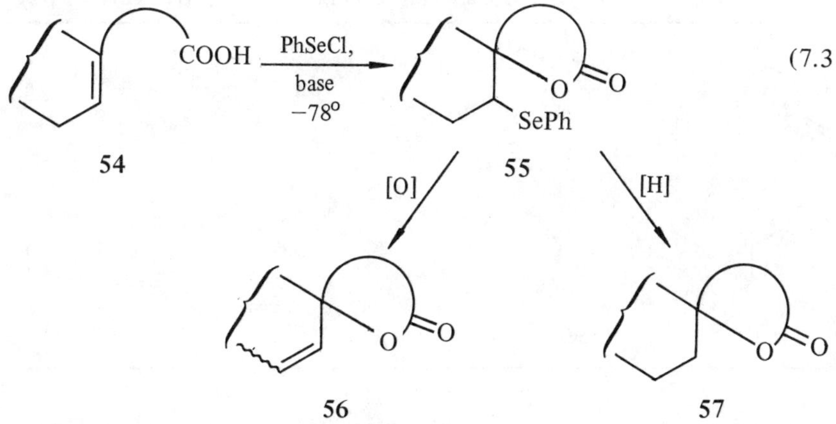

COOH —PhSeCl, base, −78°→ O O SePh (7.31)

54 **55**

[O] [H]

56 **57**

Similar results were obtained by replacing PhSeCl with phenylsulphenyl chloride, PhSCl (*phenylsulphenolactonization*).[35, 38]

The phenylselenolactonization reaction proceeds with some regio- or ring-selectivity. In general, 5-membered ring lactones are formed preferably over 4- or 6-membered rings, while 6-membered rings are preferred over 7-membered rings and 7-membered rings are preferred over 8-membered rings.[35] Table 7.2 contains some illustrative examples of this selectivity. Further examples are shown in Eq. 7.32,[34, 35] 7.33,[34–37] 7.34,[37] 7.35[39] and 7.35a.[67]

This methodology was used in a recent approach to megaphone.[68]

Table 7.2 Ring selectivity of the phenylselenolactonization reaction.

Substrate	*Phenylselenolactone*	*Yield (percent)*	*Selectivity*	*Ref.*
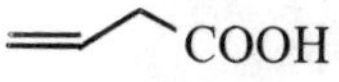 COOH	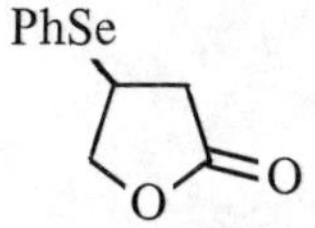 PhSe, O, O	30–35	5 *vs* 4	35, 39
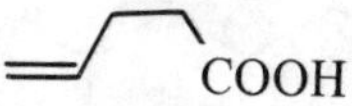 COOH	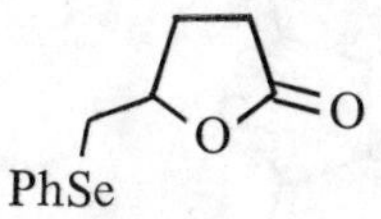 O, O, PhSe	80–100	5 *vs* 6	35–37, 4

Table 7.2–*continued*

Substrate	*Phenylselenolactone*	*Yield (percent)*	*Selectivity*	*Ref.*
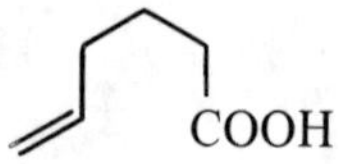 COOH	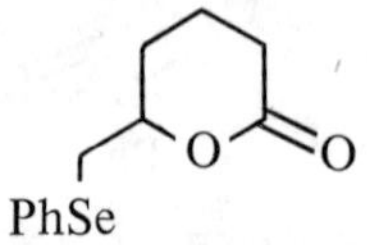 PhSe, O, O	74	6 *vs* 7	35
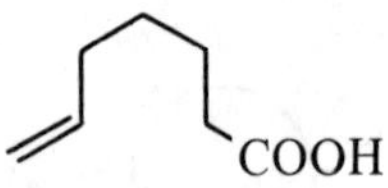 COOH	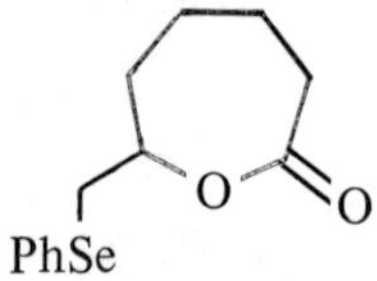 PhSe, O, O	70	7 *vs* 8	35

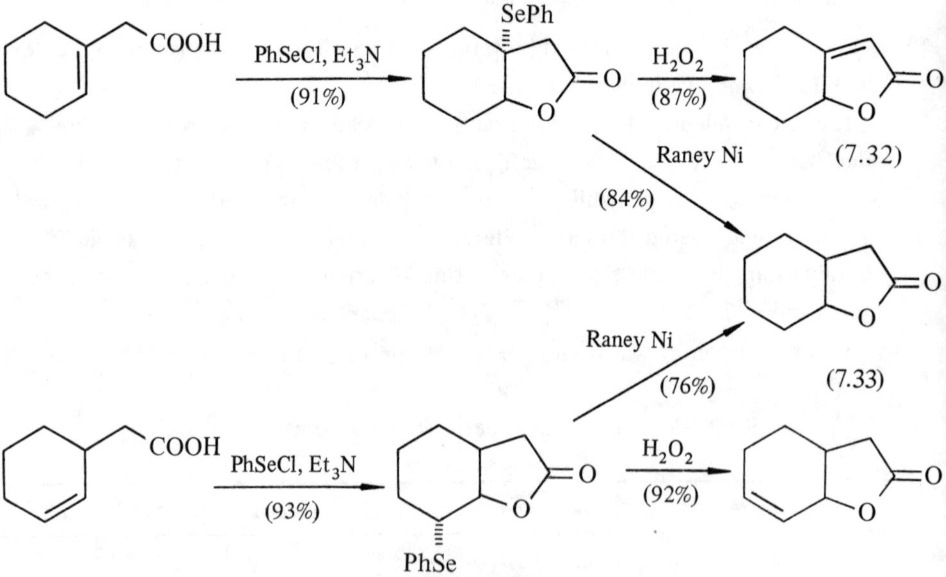

COOH —PhSeCl, Et_3N (91%)→ SePh, O, O —H_2O_2 (87%)→ O, O

Raney Ni (84%) → O, O (7.32)

COOH —PhSeCl, Et_3N (93%)→ PhSe, O, O —H_2O_2 (92%)→ O, O

Raney Ni (76%) → O, O (7.33)

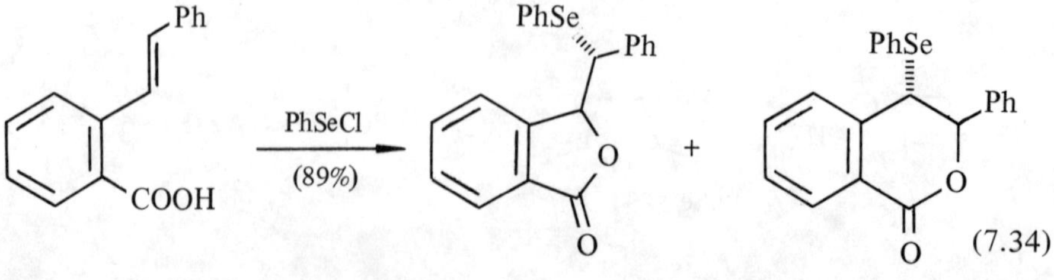

Ph, COOH —PhSeCl (89%)→ PhSe, Ph, O, O + PhSe, Ph, O, O (7.34)

Et–CH=CH–CH$_2$COOH $\xrightarrow[(95\%)]{PhSeCl,\ Et_3N}$ PhSe / Et / O / O $\xrightarrow[(93\%)]{H_2O_2}$ Et / O / O (7.35)

COOH / O / OMe $\xrightarrow[(76\%)]{PhSeCl,\ Et_3N}$ SePh / O / O / O / OMe (7.35a)

An interesting implication of the regioselectivity of the reaction is its potential for use as a means of intramolecular functionalization of an olefinic bond, as illustrated with arachidonic acid (**58**) in Eq. 7.36.[40]

COOH / **58** — PhSeCl → O / PhSe / O — (1) LiOH (2) CH_2N_2 → PhSe / OH / COOMe

a: mCPBA, THF, −78° C; A–B, 90:10

b: H_2O_2–KOH, THF, H_2O; A–B, 20:80

a or *b* (60%) → OH / COOMe (A) + HO / COOMe (B) (7.36)

Several functional groups, such as tetrahydropyranyl (THP), *t*-butyldimethylsilyl and dithiane, survive the reaction. The dithiane group, however, is reductively removed if Raney nickel is used for the removal of the PhSe group. This occurred in the synthesis of prostaglandin intermediate **59** (Eq. 7.37).[34, 35]

HOOC, S, S, $-OSi^tBuMe_2$ — PhSeCl, Et_3N (92%) → PhSe, O, O, S, S, $-OSi^tBuMe_2$

1.5 eq. H_2O_2 (86%) → O, O, S, S, $-OSi^tBuMe_2$

(70%) Raney, Ni → O, O, $-OSi^tBuMe_2$ **59** (7.37)

A procedure for the preparation of macrolides based on the phenylselenolactonization reaction was developed by Nicolaou.[4] It involves the use of N-PSS or N-PSP at 25° C in the presence of catalytic amounts of camphorsulfonic acid (CSA) under anhydrous conditions. An example is shown in Eq. 7.38.[4] Other PhSeX reagents do not lead to macrolides but only to addition products of PhSeX to the olefinic bond.[4]

OH, O — N–PSP, CSA, 25° C (50%) → O, O, SePh

n-Bu_3SnH, AIBN, Δ (100%) → O, O, Me (7.38)

In the synthesis of the naturally occurring mahubanolides **63** and **64**, the required olefinic carboxylic acids (**61** and **62**) were prepared with organoselenium methodology from α-phenylselenoester **60** (Eq. 7.39).[41] Interestingly, the phenylselenolactonization step gave selectively the lactones having the methyl and hydroxyl groups *cis*.

$R = (CH_2)_{14}Me$

(1) LDA; (2) $CH_2{=}CHCHO$; (3) H_2O_2; (4) KOH

(1) PhSeCl; (2) n-Bu_3SnH (45%)

60 **61** **62** **63** **64**

(7.39)

A one-step electrochemical phenylselenolactonization, followed by electrochemical oxidation and elimination of "PhSeOH," was reported recently.[42] Using this method, carboxylic acid **65** was converted to dihydroactinidiolide (**66**, Eq. 7.40) in high yield.[42]

PhSeSePh, Et_4NBr; Na_2CO_3, electrolysis (92%)

65 **66**

(7.40)

According to Nicolaou[23] and Clive,[24] reversal of the phenylselenolactonization is similar to the reversal of the phenylselenoetherification reaction.

During the reductive cleavage of the PhSe group of a phenylselenolactone by n-Bu_3SnH–AIBN or Ph_3SnH (§5.1), the radical generated can be trapped with an acrylate ester, resulting in an overall substitution of the PhSe group by an acrylate moiety.[43] An example of this *reductive alkylation* is given in Eq. 7.41.[43]

Ph_3SnH, 110° C
$CH_2{=}CHCOOMe$
(70%)

PhSe

COOMe

(7.41)

The PhSe group of a phenylselenolactone can also be displaced by an allyl group with the use of allyl(tri-n-butyl)stannane, as shown in Eq. 7.41a[69]

Ph Ph O O SePh

n-$Bu_3SnCH_2CH{=}CH_2$,
AIBN
(84%)

Ph Ph O O

(7.41a)

Intramolecular capture of a radical generated in similar fashion from the addition of phenylselenenyl chloride and silver acrylate or crotonate to an olefin, results in the formation of a γ-lactone, as illustrated in Eq. 7.42.[44]

PhSeCl
MeCH=CHCOOAg
(74%)

SePh O O

Ph_3SnH,
AIBN, Δ
(61%)

H H O O

(7.42)

7.3 Synthesis of S-heterocycles

Extension of the organoselenium-mediated cyclizations to olefinic thio compounds (**67**) leads to the formation of S-heterocycles.[3,5,45] Reductive removal of the PhSe group from the phenylselenothioethers (**68**), to give thioethers (**69**) is readily achieved by tri-n-butyltin hydride at high temperature in the presence of the radical initiator AIBN (Eq. 7.43). Oxidative treatment of **68**, however, gives a different type of product, depending on the oxidation method used (Eq.

7.44). The products (**70**, **71** and **72**) are the result of oxidation of the heterocyclic sulfur atom to the sulfoxide or sulfone, after or prior to the selenoxide elimination.

PhSeCl

RS

67

R = H, Ac

S

SePh

68

n-Bu_3SnH

AIBN, Δ

S

69

(7.43)

8 eq. H_2O_2

3 eq. mCPBA, −78° C

(1) 1 eq. mCPBA, −78° C
(2) 1 eq. mCPBA, −20° C

S O O

70

S O O

71

S O

72

(7.44)

Nicolaou has applied the method to the synthesis of S-containing analogs of prostacyclin such as **73**, **74** and **75** (Eq. 7.45).[45,46]

AcS

COOMe

OSi^tBuMe_2

OSi^tBuMe_2

(7.45)

PhSeCl

(62%)

COOMe

SePh

S

OSi^tBuMe_2

OSi^tBuMe_2

(94%) n-Bu_4NF

COOMe

SePh

2 eq. mCPBA

(71%)

COOMe

73

8 eq. H_2O_2

(75%)

(22%)

3 eq. mCPBA

(86%)

74

75

An example of a different type of selenium-mediated cyclization to form an S-heterocycle was reported by Baldwin and is presented in Eq. 7.46.[47]

PhSeBr, pyridine

(1) mCPBA

(2) NH_3

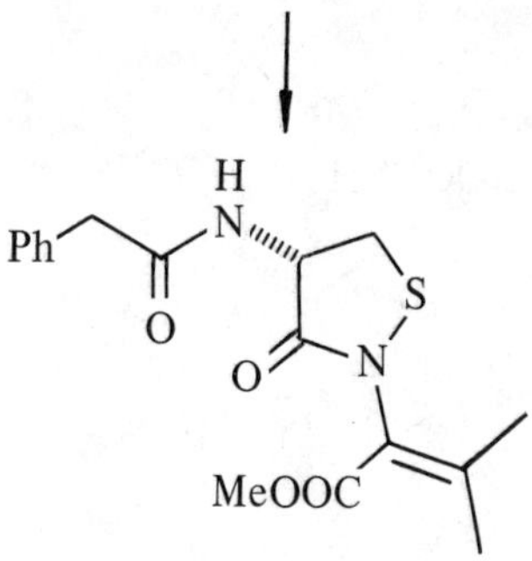
H
N
Ph
O
S
O
N
MeOOC

(7.46)

7.4 *Synthesis of N-heterocycles*

Nitrogen heterocycles were prepared by selenium-induced cyclizations of olefinic urethanes. The method has potential in the synthesis of alkaloids. Some examples are shown in Eq. 7.47,[48] 7.48,[48] 7.49[4] and 7.50.[49]

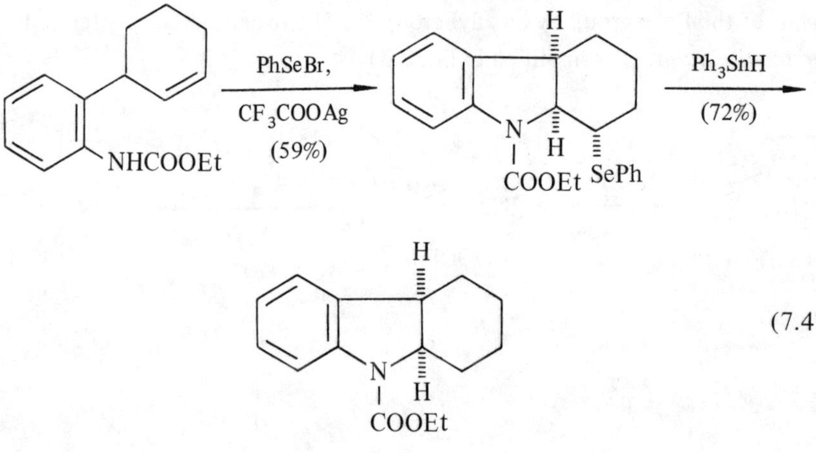
NHCOOEt
PhSeBr,
CF3COOAg
(59%)
H
N
H
COOEt
SePh
Ph3SnH
(72%)
H
N
H
COOEt

(7.47)

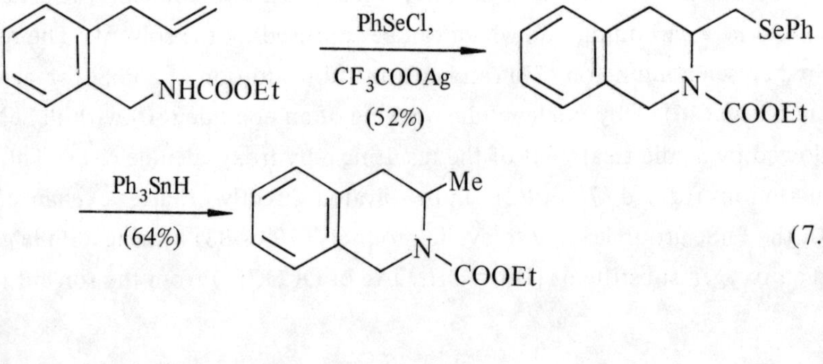
NHCOOEt
PhSeCl,
CF3COOAg
(52%)
SePh
N
COOEt
Ph3SnH
(64%)
Me
N
COOEt

(7.48)

EtOOCNH, Me; N–PSP (90%); EtOOC, N, Me, SePh (7.49)

NH; PhSeBr (63%); H, SePh, N (7.50)

Treatment of phenylselenourethanes, prepared as above with allyl(tri-*n*-butyl)-stannane in the presence of azobisisobutyronitrile AIBN), leads to the replacement of the PhSe group by an allyl group.[50] The overall sequence, termed *ureidoallylation*, is exemplified in Eq. 7.51.[50]

NH, $COOCH_2Ph$; N–PSP; N, SePh, $COOCH_2Ph$; $n\text{-}Bu_3Sn$, AIBN (70%); N, $COOCH_2Ph$

(7.51)

7.5 Synthesis of carbocycles

Intramolecular capture of an episelenonium ion by an olefinic bond results in the formation of a new ring via a new carbon–carbon bond and the generation of a new carbonium ion which can be captured by the solvent. The required episelenonium ion (**77**) can be formed by addition of a PhSeX reagent to an olefin (**76**) or by nucleophilic opening of an epoxide (**80**) with PhSeNa, followed by acidic treatment of the resulting β-hydroxyselenide (**81**). The carbonium ions formed (**78** or **82**) can be solvated directly or, after exchange, with the PhSe group leading to cyclic products (**79** or **83**) bearing a PhSe group and an oxygen substituent (OR = OH, OAc or $OCOCF_3$) from the solvent (Eq.

7.52). The PhSe group can then be eliminated oxidatively (§4.2) or reductively (§5.1) to give the corresponding selenium-free carbocycles.

(7.52)

The first example of this type of selenium-induced carbocyclization was observed by Clive in the conversion of cyclic diolefin 84 to bicyclic alcohol 85 (Eq. 7.53).[51]

(7.53)

Kametani has used the β-hydroxyselenides (**81**) as an entry to this sequence and has applied the methodology to the synthesis of safranal (**87**) from geranyl acetate (**86**, Eq. 7.54)[52] and *p*-menthan derivatives **89** and **90** from linanyl acetate (**88**, Eq. 7.55).[53] These syntheses were considered to be biomimetic, since they resemble the biosynthesis of these compounds.

(7.54)

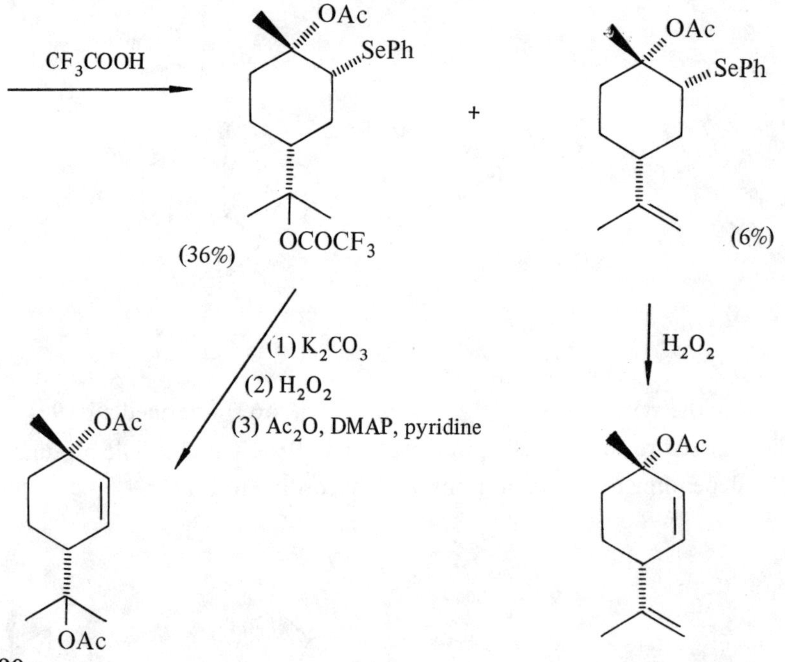

CF_3COOH

OAc SePh (36%) $OCOCF_3$ + OAc SePh (6%)

(1) K_2CO_3
(2) H_2O_2
(3) Ac_2O, DMAP, pyridine

H_2O_2

OAc OAc **90**

OAc **89** (7.55)

The presence of a second internal nucleophile such as hydroxyl, carboxyl, etc., may result in the formation of a second ring by intramolecular capture of the carbonium ion. This type of transformation allowed the conversion of nerolidol (**91**) to caparrapi oxide (**92**) and its isomer (Eq. 7.56).[54] Similar transformations were also observed by French workers.[55]

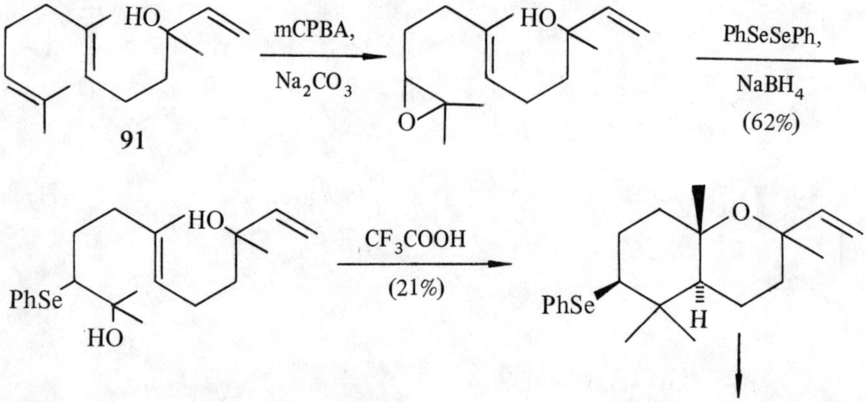

HO **91** mCPBA, Na_2CO_3 HO O PhSeSePh, $NaBH_4$ (62%)

HO PhSe HO CF_3COOH (21%) O PhSe H

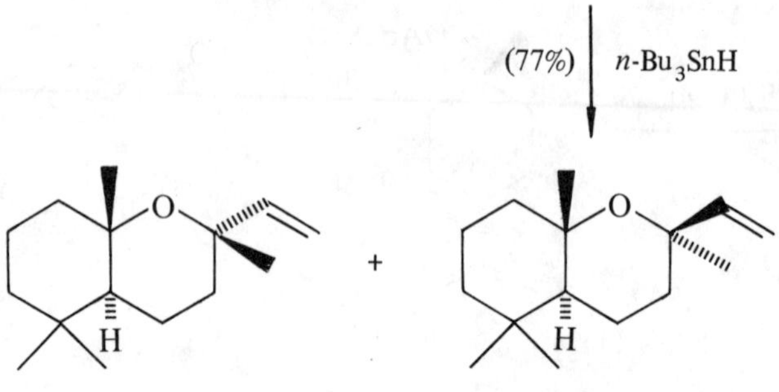

(77%) n-Bu_3SnH

H

92

+

H

(7.56)

During the synthesis of aplysistatin (**121**, Eq. 4.63) intermediate (**93**) could be cyclized to the tricyclic products **94** and **95**, or the bicyclic products **96** and **97**, depending on the conditions used, according to Eq. 7.57.

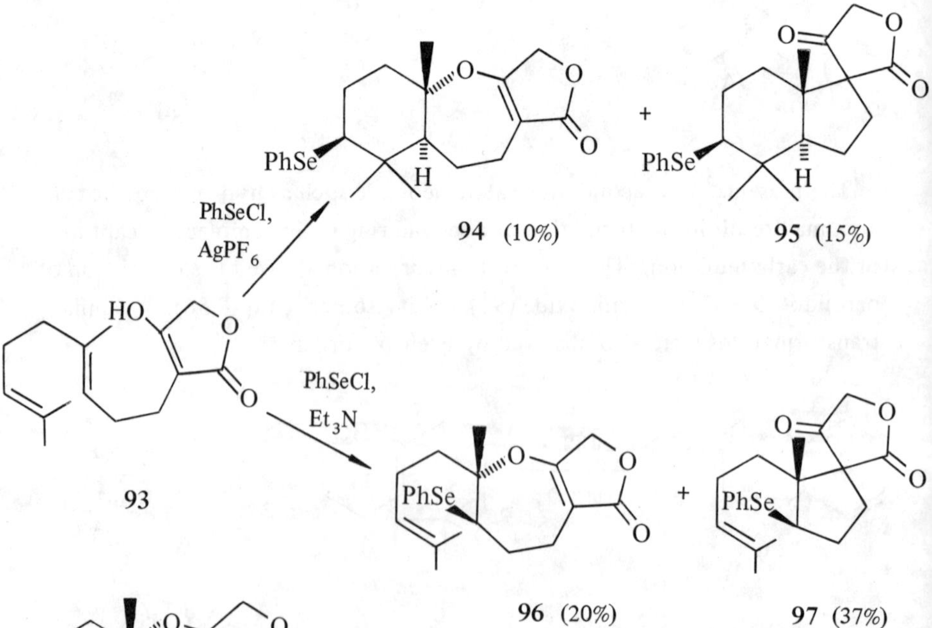

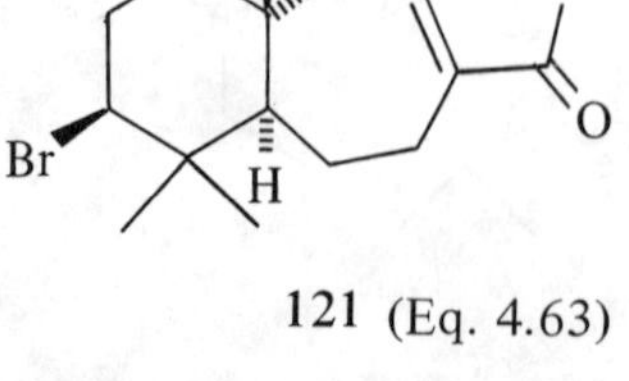

PhSeCl, $AgPF_6$

PhSeCl, Et_3N

HO

93

PhSe H **94** (10%)

PhSe H **95** (15%)

PhSe **96** (20%)

PhSe **97** (37%)

Br H

121 (Eq. 4.63)

(7.57)

Stepwise phenylselenolactonization and carbocyclization was necessary in order to convert homogeranic acid (**98**) to the bicyclic lactone (**100**, Eq. 7.58)[57]. The carbocyclization step was induced by an episelenonium ion, formed from the intermediate phenylselenolactone (**99**) with assistance from the lactone functionality.

COOH — N–PSP or PhSeCl → PhSe, =O

98 **99**

(7.58)

CH_3SO_3H or H_2SO_4 → PhSe, H, =O

100

A carbocyclization reaction induced by "PhSeI" was recently reported by Toshimitsu[58]. This reagent is prepared *in situ* from PhSeSePh and iodine, and the cyclization products bear the MeCONH– group which arises from participation of MeCN used as solvent. An example is shown in Eq. 7.59[58].

PhSeSePh, I_2, MeCN (75%) → PhSe, NHCOMe

(7.59)

+ 83 : 17

PhSe, NHCOMe

Se-aryl arene selenosulfonates ($ArSO_2SePh$) react similarly, as indicated in Eq. 7.60[59].

$$\xrightarrow[(34\%)]{ArSO_2SePh,\ h\nu}\qquad\xrightarrow[(2)\ H_2,\ Pd/C]{(1)\ H_2O_2}$$

H SO$_2$Ar H SO$_2$Ar

PhSe H H

(Ar = *p*-Tolyl) (7.60)

Enolic-type olefinic bonds were also found to participate in selenium-induced carbocyclizations. Thus, Ley[60—63] has developed methodology for the synthesis of carbocycles (**102**) from olefinic β-ketoesters (**101**)[60, 61] or olefinic α-phenylseleno-β-ketoesters (**103**)[62], as summarized in Eq. 7.61.

O COOR PhSeX O COOR

SePh

101 **102**

SnCl$_4$

COOR

O COOR SePh Acid O (7.61)

SePh

103 **104**

Participation of the enolic oxygen in the cyclization results in the formation of O-heterocycles (**104**) as by-products or even as the main products (§7.1). These are usually the kinetically favored products and they can be converted to the carbocycles (**102**) by treatment with SnCl$_4$. Some typical examples of these cyclizations are shown in Eqs. 7.62[60, 61], 7.63[61], 7.64[61], and 7.65[62].

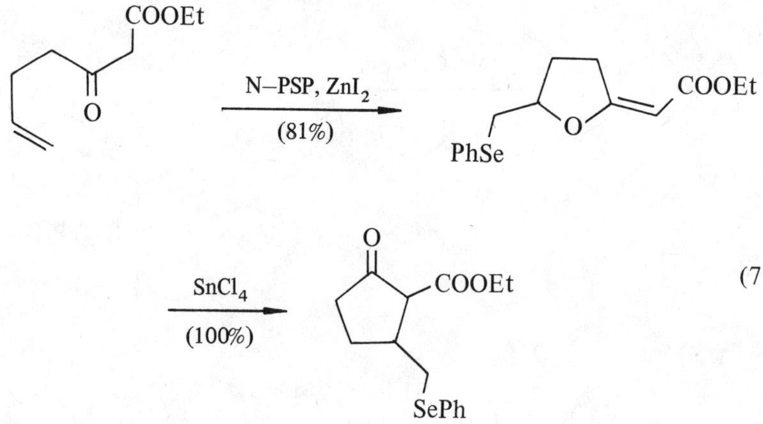

N–PSP, ZnI_2 (81%) ; $SnCl_4$ (100%)

(7.62)

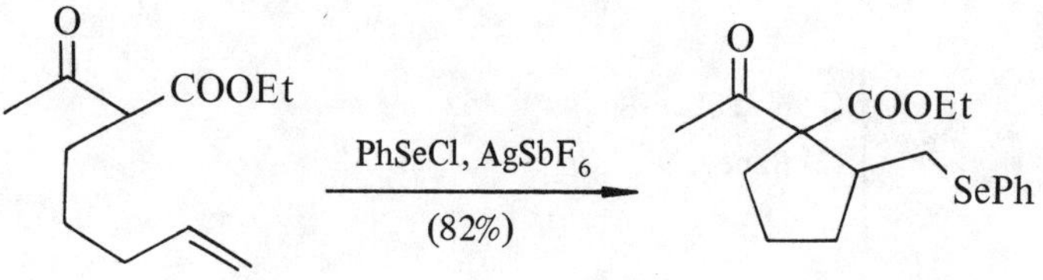

PhSeCl, $AgSbF_6$ (82%)

(7.63)

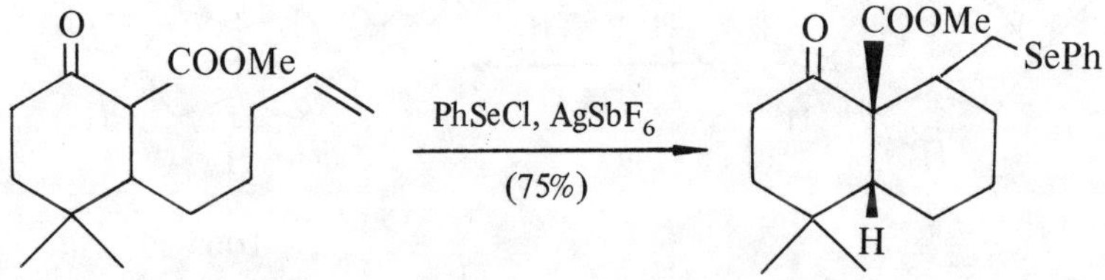

PhSeCl, $AgSbF_6$ (75%)

(7.64)

$SnCl_4$

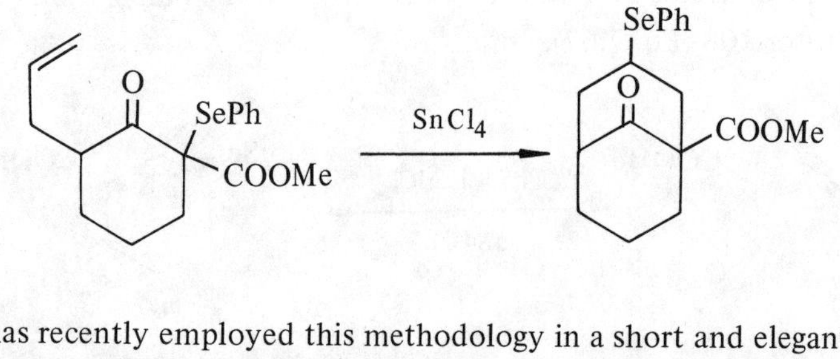

(7.65)

Ley has recently employed this methodology in a short and elegant synthesis of hirsutene (**105**, Eq. 7.66)[63].

(1) Mg, CuBr · Me_2S
(2) COOMe
(74%)
MeOOC
(90%) N–PSP, $SnCl_4$
(1) Raney, Ni
(2) LDA
(3) $LiAH_4$
(65%)
(52%) (1) N–PSP, n-Bu_3P
(2) Raney, Ni
$Ph_3P{=}CH_2$
(70%)
106

(7.66)

A similar selenium-induced carbocyclization was also reported by Weiler[64], who used the derived carbocyclic product (**107**) for the preparation of carotenoid precursor **108** (Eq. 7.67).

COOMe
PhSeCl, $AlCl_3$
(84%)
PhSe COOMe
107

(1) Base
(2) $ClP(O)(OEt)_2$

PhSe COOMe LiCuMe$_2$ PhSe COOMe $OP(O)(OEt)_2$

[O]

COOMe

108

(7.67)

A different kind of carbocyclization reaction, based on vinyl selenides, was recently reported and is shown in Eq. 7.68[65].

Cl O (1) PhSeNa (2) NaH (3) MsCl, LiCl Cl PhSe O COOMe (65%)

SePh O COOMe (1) CF_3COOH (2) H_2O_2 (63%) O COOMe

(7.68)

Organotin derivatives can also trap intramolecularly an episelenonium intermediate leading again to carbocycles. Nicolaou[4] has reported an example of synthesis of a cyclopropane (**110**) using olefinic organotin compound **109** and N–PSP, as indicated in Eq. 7.69.

$$\text{109 (SnMe}_3\text{)} \xrightarrow{\text{N–PSP}} \text{110 (PhSe)} \qquad (7.69)$$

A free radical carbocyclization, initiated from a phenyl selenide, was observed during the treatment of **111** with $n\text{-Bu}_3\text{SnH}$[66]. The cyclized product (**112**) was obtained together with the "normal" product of the reductive elimination of the selenide (**113**, Eq. 7.70)[66].

$$\text{(Cl)} \xrightarrow[\text{NaOH, } n\text{-Bu}_4\text{NBr (65\%)}]{\text{PhSeH}} \text{111 (SePh)} \qquad (7.70)$$

$$\xrightarrow[\text{AIBN, } 80^\circ\text{ C}]{n\text{-Bu}_3\text{SnH,}} \text{112 (38\%)} + \text{113 (Me) (27\%)}$$

REFERENCES

1. Clive, D. L. J., Chittattu, G., Curtis, N. J., Kiel, W. A., and Wong, C. K., *J. C. S. Chem. Comm.*, 725 (1977).
2. Nicolaou, K. C., and Lysenko, Z., *Tetrahedron Lett.*, 1257 (1977).
3. Nicolaou, K. C., Magolda, R. L., Sipio, W. J., Barnette, W. E., Lysenko, Z.,

and Joullié, M. M., *J. Am. Chem. Soc.,* **102**, 3784 (1980).

4. Nicolaou, K. C., Claremon, D. A., Barnette, W. E., and Seitz, S. P., *J. Am. Chem. Soc.,* **101**, 3704 (1979).
5. Nicolaou, K. C., *Tetrahedron,* **37**, 4097 (1981).
6. Clive, D. L. J., Chittattu, G., and Wong, C. K., *Can. J. Chem.,* **55**, 3894 (1977).
7. (a) Corey, E. J.,Keck,G. E., and Székely, I., and Ishigura, M., *Tetrahedron Lett.,* 1023 (1978).
8. (a) Nicolaou, K. C., Barnette, W. E., and Magolda, R. L., *J. Am. Chem. Soc.,* **103**, 3480 (1981); (b) Nicolaou, K. C., and Barnette, W. E., *J. C. S. Chem. Comm.,* 331 (1977).
9. Lysenko, Z., Ricciardi, F., Semple, J. E., Wang, P. C., and Joullié, M. M., *Tetrahedron Lett.,* 2679 (1978).
10. Hoye, T. R., and Caruzo, A. J., *J. Org. Chem.,* **46**, 1198 (1981).
11. Kraus, G. A., and Taschner, M. J., *J. Org. Chem.,* **45**, 1175 (1980).
12. Kane, P. D., and Mann, J., *J. C. S. Chem. Comm.,* 224 (1983).
13. Narutz, Y., Hidemitsu, U., and Muruyama, K., *J. C. S. Chem. Comm.,* 1277 (1981).
14. Fukuyama, T., Robins, B. D., and Sachleben, R. A., *Tetrahedron Lett.,* 22, 4155 (1981).
15. Current, S., and Sharpless, K. B., *Tetrahedron Lett.,* 5075 (1978).
16. Stevens, R. V., and Hrib, N., *Tetrahedron Lett.,* 22, 4791 (1981).
17. Ley, S. V., and Lygo, B., *Tetrahedron Lett.,* 23, 4625 (1982).
18. Ley, S. V., Lygo, B., Molines, H., and Morton, J. A., *J. C. S. Chem. Comm.,* 1251 (1982).
19. Petrzilka, M., *Helv. Chim. Acta,* **61**, 3075 (1978).
20. Kraus, G. A., and Roth, B., *J. Org. Chem.,* **45**, 4825 (1980).
21. Nicolaou, K. C., Sipio, W. J., Magolda, R. L., Seitz, S., and Barnette, W. E., *J. C. S. Chem. Comm.,* 1067 (1978).
22. Nicolaou, K. C., and Petasis, N. A., unpublished results.
23. Nicolaou, K. C., Sipio, W. J., Magolda, R. L., and Claremon, D. A., *J. C. S. Chem. Comm.,* 83 (1979).
24. Clive, D. L. J., and Kale, V. N., *J. Org. Chem.,* **46**, 231 (1981).
25. Nicolaou, K. C., and Webber, S., unpublished results.
26. Beaulieu, P. L., Morisset, V. M., and Garratt, D. G., *Tetrahedron Lett.,* **21**, 129 (1980).
27. Filer, C. N., Ahern, D., Fazio, R., and Shelton, E. J., *J. Org. Chem.,* **45**, 1313

(1980).

28. Scarborough, R. M., Jr., Smith, A. B., III, Barnette, W. E., and Nicolaou, K. C., *J. Org. Chem.,* **44**, 1742 (1979).
29. Toshimitsu, A., Aoai, T., Uemura, S., and Okano, M., *J. Org. Chem.,* **46**, 3021 (1981); *J. C. S. Chem. Comm.,* 610 (1979); *Chem. Lett.,* 1359 (1979).
30. Uemura, S., Toshimitsu, A., Aoai, T., and Okano, M., *Tetrahedron Lett.,* **21**, 1533 (1980).
31. Kametani, T., Nemoto, H., and Fukumoto, K., *Bioorganic Chem.,* **7**, 215 (1978); *Heterocycles,* **6**, 1365 (1977).
32. Taurand, G., Beau, J.-M., and Sinay, P., *J. C. S. Chem. Comm.,* 701 (1982).
33. (a) Nicolaou, K. C., *Tetrahedron,* **33**, 683 (1977); (b) Masamune, S., Bates, G. S., and Corcoran, J. W., *Angew. Chem. Int. Ed. Engl.,* **16**, 587 (1977); (c) Back, T. G., *Tetrahedron,* **33**, 3041 (1977).
34. Nicolaou, K. C., and Lysenko, Z., *J. Am. Chem. Soc.,* **99**, 3185 (1977).
35. Nicolaou, K. C., Seitz, S. P., Sipio, W. J., and Blount, J. F., *J. Am. Chem. Soc.,* **101**, 3884 (1979).
36. Clive, D. L. J., and Chittattu, G., *J. C. S. Chem. Comm.,* 484 (1977).
37. Clive, D. L. J., Russell, C. G., Chittattu, C., and Singh, A., *Tetrahedron,* **36**, 1399 (1980).
38. Nicolaou, K. C., and Lysenko, Z., *J. C. S. Chem. Comm.,* 293 (1977).
39. Goldsmith, D., Liotta, D., Lee, C., and Zima, G., *Tetrahedron Lett.,* 4801 (1979).
40. Baldwin, J. E., Reed, N. V., and Thomas, E. J., *Tetrahedron,* **37**, 263 (1981).
41. Rollinson, W. W., Amos, R. A., and Katzenellenbogen, J. A., *J. Am. Chem. Soc.,* **103**, 4114 (1981).
42. Torii, S., Uenyama, K., Ono, M., and Bannon, T., *J. Am. Chem. Soc.,* **103**, 4606 (1981).
43. Burke, S. D., Fobare, W. F., and Armistead, D. M., *J. Org. Chem.,* **47**, 3348 (1982).
44. Clive, D. L. J., and Beaulieu, P. L., *J. C. S. Chem. Comm.,* 307 (1983).
45. Nicolaou, K. C., Barnette, W. E., and Magolda, R. L., *J. Am. Chem. Soc.,* **100**, 2567 (1978).
46. Nicolaou, K. C., Barnette, W. E., and Magolda, R. L., *J. Am. Chem. Soc.,* **103**, 3486 (1981).
47. Baldwin, J. E., Haber, S. B., Kitchin, J., *J. C. S. Chem. Comm.,* 790 (1973).
48. (a) Clive, D. L. J., Farina, V., Singh, A., Wong, C. K., Kiel, W. A., and Menchen, S. M., *J. Org. Chem.,* **45**, 2120 (1980); (b) Clive, D. L. J., Wong, C. K.,

Kiel, W. A., and Menchen, S. M., *J. C. S. Chem. Comm.*, 379 (1978).

49. Wilson, S. R., and Sawicki, R. A., *J. Org. Chem.*, **44**, 287 (1979).
50. Webb, R. R., II, and Danishefsky, S., *Tetrahedron Lett.*, 1357 (1983).
51. Clive, D. L. J., Chittattu, G., and Wong, C. K., *J. C. S. Chem. Comm.*, 441 (1978).
52. Kametani, T., Suzuki, K., Kurobe, H., and Nemoto, H., *Chem. Pharm. Bull.*, **29**, 105 (1981); *J. C. S. Chem. Comm.*, 1128 (1979).
53. Kametani, T., Kurobe, H., and Nemoto, H., *J. C. S. Perkin I*, 756 (1981); *J. C. S. Chem. Comm.*, 762 (1980).
54. Kametani, T., Fukumoto, K., Kurobe, H., and Nemoto, H., *Tetrahedron Lett.*, 3653 (1981).
55. Perrier, M., and Rouessac, F., *Compt. rend. Acad. Sci., Ser. II*, **295**, 729 (1982).
56. White, J. D., Nishiguchi, T., and Skeean, R. W., *J. Am. Chem. Soc.*, **104**, 3923 (1982).
57. (a) Rouessac, A., Rouessac, F., and Zamarlik, H., *Tetrahedron Lett.*, 22, 2641 (1981); (b) Rouessac, F., and Zamarlik, H., *Tetrahedron Lett.*, 22, 2643 (1981).
58. Toshimitsu, A., Uemura, S., and Okano, M., *J. C. S. Chem. Comm.*, 87 (1982).
59. Gancarz, R. A., and Kice, J. J., *J. Org. Chem.*, **46**, 4899 (1981).
60. Jackson, W. P., Ley, S. V., and Morton, J. A., *J. C. S. Chem. Comm.*, 1028 (1980).
61. Jackson, W. P., Ley, S. V., and Whittle, A. J., *J. C. S. Chem. Comm.*, 1173 (1980).
62. Jackson, W. P., Ley, S. V., and Morton, J. A., *Tetrahedron Lett.*, 22, 2601 (1981).
63. Ley, S. P., and Murray, P. J., *J. C. S. Chem. Comm.*, 1252 (1982).
64. Alderdice, M., and Weiler, L., *Can. J. Chem.*, **59**, 2239 (1981).
65. Gravel, D., Déziel, R., and Bordeleau, L., *Tetrahedron Lett.*, **24**, 699 (1983).
66. (a) Bachi, M. D., Frolow, F., and Hoornaert, C., *J. Org. Chem.*, **48**, 1841 (1983); (b) Bachi, M. D., and Hoornaert, C., *Tetrahedron Lett.*, 22, 2693 (1981).
67. Burke, S. D., Fobare, W. F., and Pacofsky, G. I., *J. Org. Chem.*, **48**, 5221 (1983).
68. Hoye, T. R., and Kurth, M. J., *Tetrahedron Lett.*, **24**, 4769 (1983).
69. Keck, G. E., and Yates, J. B., *J. Am. Chem. Soc.*, **104**, 5829 (1982).

CHAPTER 8

Miscellaneous Applications of Selenium Compounds

The diversity and versatility of organoselenium chemistry has led to the discovery of some more applications of selenium compounds, other than the ones categorized in Chapters 2–7. Several of these novel and potentially useful methods are briefly presented in this chapter.

8.1 Synthesis and reactions of selenol esters

Selenol esters (2) have been prepared from the corresponding acids (1) by treatment with phenyl selenocyanate (PhSeCN)[1] or, more conveniently, with *N*-phenylselenophthalimide (N-PSP)[2] and tri-*n*-butylphosphine (n-Bu_3P), as in Eq. 8.1. Examples are shown in Eq. 8.2[1] and 8.3.[2]

$$\underset{\mathbf{1}}{R-\overset{O}{\overset{\|}{C}}-OH} \xrightarrow[\text{or N–PSP, } n\text{-Bu}_3\text{P}]{\text{PhSeCN, } n\text{-Bu}_3\text{P}} \underset{\mathbf{2}}{R-\overset{O}{\overset{\|}{C}}-SePh} \quad (8.1)$$

$$\text{cyclohexenyl–COOH} \xrightarrow[(84\%)]{\text{PhSeCN, } n\text{-Bu}_3\text{P}} \text{cyclohexenyl–COSePh} \quad (8.2)$$

$$\text{HOOC–}CH_2\text{–(OMe furanoside acetonide)} \xrightarrow[(94\%)]{\text{N–PSP, } n\text{-Bu}_3\text{P}} \text{PhSe–}\overset{O}{\overset{\|}{C}}\text{–}CH_2\text{–(OMe furanoside acetonide)} \quad (8.3)$$

Various carboxylic acid derivatives were also converted to selenol esters upon treatment with the appropriate selenium reagents. These include *N*-acylhydrazines (**3**), which are treated with benzeneseleninic acid, as in Eq. 8.4,[3] and carboxylic esters (**4**) or lactones, which are reacted with $Me_2AlSeMe$ (prepared from Me_3Al and Se) (Eq. 8.5).[4] Typical examples are shown in Eq. 8.6,[3] 8.7[4] and 8.8.[4]

$$\underset{\mathbf{3}}{R-\overset{O}{\overset{\|}{C}}-NHNH_2} \xrightarrow{PhSeO_2H,\ Ph_3P} \underset{\mathbf{2}}{R-\overset{O}{\overset{\|}{C}}-SePh} \tag{8.4}$$

$$\underset{\mathbf{4}}{R-\overset{O}{\overset{\|}{C}}-OR'} \xrightarrow{Me_2AlSeMe} \underset{\mathbf{5}}{R-\overset{O}{\overset{\|}{C}}-SeMe} \tag{8.5}$$

OMe … CONHNH$_2$ $\xrightarrow[(84\%)]{PhSeO_2H,\ Ph_3P}$ OMe … COSePh (8.6)

$$\text{(cyclopropyl)}-COOEt \xrightarrow[(96\%)]{Me_2AlSeMe} \text{(cyclopropyl)}-COSeMe \tag{8.7}$$

O, =O (lactone) $\xrightarrow[(80\%)]{Me_2AlSeMe}$ OH, COSeMe (8.8)

Selenol esters (**2** or **5**) are useful acylating agents. They have been used to prepare amides (**6**, Eq. 8.9),[2] ketones (**7**, Eq. 8.10[4] and **8**, Eq. 8.11[4]), carboxylic acids (**9**, Eq. 8.11),[4] esters (**10**, Eq. 8.11)[4] and isoxazoles (**11**, Eq. 8.12).[4] They also react with organocuprates to form ketones.[116]

COOH, OMe → N–PSP, n-Bu_3P, i-$PrNH_2$ (95%) → CONHiPr, OMe **6** (8.9)

SeMe, O → $(CuOTf)_2 \cdot PhH$ (70%) → O **7** (8.10)

SeMe, O → $(CuOTf)_2 \cdot PhH$, furan (100%) → O **8**

H_2O (97%) → O, OH **9**

MeOH (88%) → O, OMe **10** (8.11)

COSeMe + COOEt, N≡C → Cu_2O, Et_3N (61%) → COOEt, N, O **11** (8.12)

In situ preparation of a selenol ester by oxidation of compound **13** and subsequent formation of the corresponding selenoxide (**14**), allowed the application of heteroconjugate addition (**12**→**13**), for the synthesis of α-substituted carboxylic acids, according to Eq. 8.13.[5] Isobe employed this methodology in an approach to Prelog–Djerassi lactone (**16**, Eq. 8.14).[5]

(8.13)

SiMe$_3$ (1) RLi (2) PhSeCl SO$_2$Ph **12** → R SiMe$_3$ SePh SO$_2$Ph **13** ; R COOH **15**

H_2O_2

[R SiMe$_3$ O SePh SO$_2$Ph → R ÖSiMe$_3$ SePh SO$_2$Ph → H_2O_2 → R H_2O O Se–Ph O] **14**

(1) $(COCl)_2$, DMSO, Et_3N
(2) $PhS(Me_3Si)_2CLi$
(3) mCPBA
(60%)

EtO O OH → EtO O SO$_2$Ph SiMe$_3$

(100%) (1) MeLi (2) PhSeCl

EtO O SO$_2$Ph SePh SiMe$_3$

(1) H_2O_2
(2) HCl
(3) NaOAc, Br_2

O O COOH **16**

(8.14)

8.2 Synthesis and reactions of vinyl selenides

Vinyl selenides (**17**) are quite versatile synthetic intermediates with great potential. They are accessible by several routes (Eq. 8.15) and they can be readily converted to useful products. The various methods used to prepare this type of compound are summarized in Eq. 8.15. It should be noted, however, that the geometry of the olefinic bond varies significantly according to the method of preparation. As indicated in Eq. 8.15, common precursors for the synthesis of vinyl selenides are acetylenes (**18**),[5] acetylenic selenides (**19**),[5,6] olefins (**20**),[7] ketene selenoacetals (**21**),[8] vinylic organometallics (**22**),[6,9] α-haloselenides (**23**),[10] selenoacetals (**24**, §5.1, 6.1),[11,12] and carbonyl compounds (**25**).[5,9,13–15] Some examples are shown in Eq. 8.16,[5,6] 8.17,[7] 8.18,[9] 8.19,[9,11] 8.20[14] and 8.21.[5,15]

18: {—≡—H
23: cyclopropyl(H)–CH(H)(Br)(SePh)
24: cyclopropyl(H)–CH(H)(SePh)(SePh)

18 → 19: (1) *n*-BuLi (2) PhSeBr
18 → 17: PhSeH
23 → 17: Base or DMF
24 → 17: PI_3 or MeI, DMF

19: {—≡—SePh → 17: $LiAlH_4$ or $(C_6H_{11})_2BH$ and AcOH

17: vinyl selenide (SePh, H)

25: carbonyl (=O) → 17: $Ph_3P{=}CHSePh$ or $Ph_2P(O)\overset{\ominus}{C}HSePh$ or $(PhSe)_2CH^{\ominus}$ and MsCl, Et_3N

20 → 17: (1) PhSeBr (2) *t*-BuOK
21 → 17: (1) *n*-BuLi (2) $H_3O^{\oplus}$
22 → 17: PhSeCl or PhSeBr

(8.15)

20: olefin (H, H)
21: ketene selenoacetal (SePh, SePh)
22: vinyl organometallic (M, H)

M = MgBr, Li, HgCl, $B(OH)_2$–NaOH

Ph SePh

$(C_6H_{11})_2BH$ and AcOH (90%)

$$\text{PhC}\equiv\text{CH} \xrightarrow[\text{(2) PhSeBr}]{\text{(1) } n\text{-BuLi}} \text{PhC}\equiv\text{C—SePh} \qquad (8.16)$$

(90%)

$LiAlH_4$ (92%)

Ph SePh

(1) PhSeBr, 25° C

(2) *t*-BuOK, 25° C

(92%)

SePh

(8.17)

SePh SePh (1) LDA (2) MeCOMe (95%) HO SePh SePh H MsCl, Et_3N (77%) SePh

(8.18)

SePh SePh MeI, DMF (66%) SePh PhSeBr (67%) MgBr (8.19)

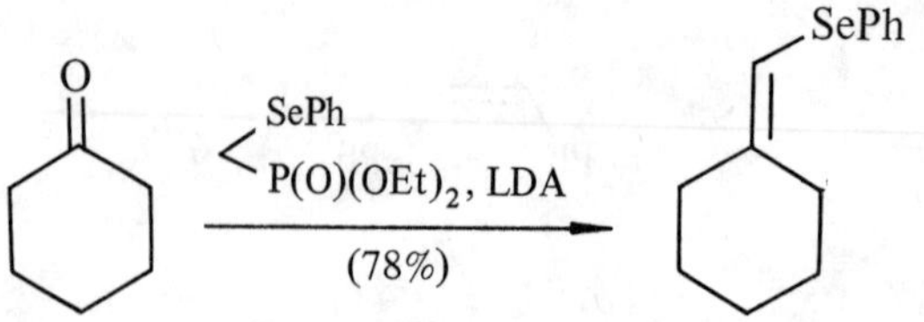

(8.20)

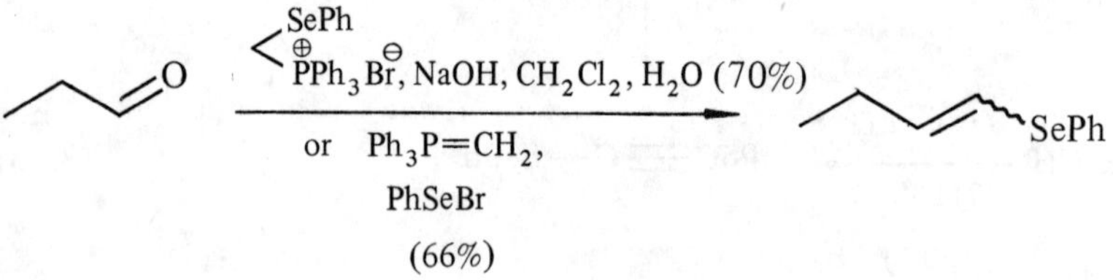

(8.21)

More highly functionalized vinyl selenides having various substituents attached to the olefinic bond, have also been made, either by alkylation of the carbanions derived from the unsubstituted vinyl selenides (Eq. 6.31), or by addition of organo-selenium electrophiles to acetylenes or enones. Some examples of this type of chemistry are given in Eq. 6.28, 6.29, 7.23, 8.22,[16] 8.23,[17] 8.24,[18] 8.25,[19] 8.26,[20] 8.27,[21] 8.28[22] and 8.29.[23]

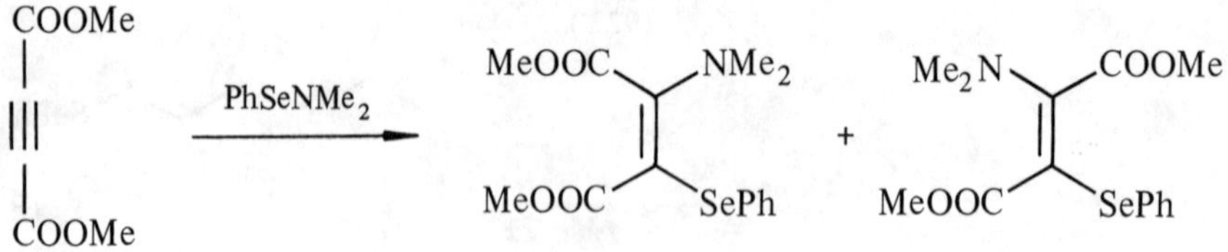

55 : 45 (8.22)

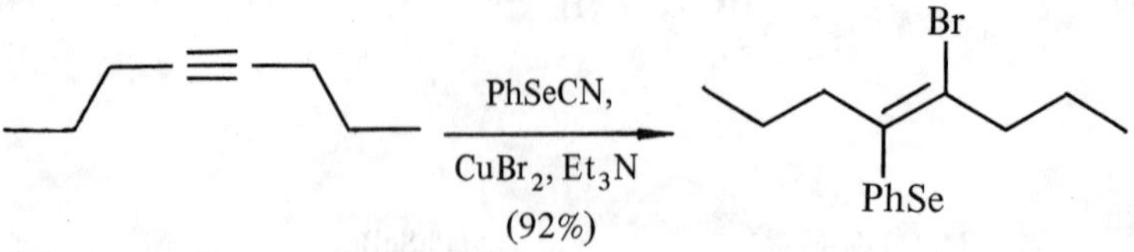

(8.23)

PhSeBr (93%) — Al, Hg (76%) — DBU (68%) — (62%) — DBU (40%)

Cl, Cl, Br, PhSe, O

(8.24)

PhSeSO₂Ar (52%) — SO₂Ar, SePh — (75%) H₂O₂ — ArSO₂

(8.25)

B + Li — B⊖ Li⊕ — PhSeCl

H_2O_2
(70%)

SePh
B
2

(79%) Me_3NO

SePh

(2) H_2O_2
(69%)

(8.26)

HO

$(p\text{-Cl-}C_6H_4Se)_2$,
Et_4NClO_4, H_2SO_4,
electrolysis
(87%)

H
SePh

(1)
(2) H_2O_2
(81%)

CHO

(8.27)

PhSeCl, pyridine
(71%)

SePh

(8.28)

(8.29)

$PhSeSiMe_3$,
cat. ZnI_2

OAc

SePh

PhSeH

Br

As already mentioned in §6.1 and 6.3, vinyl selenides can be converted to carbanions and then alkylated, or they can be transformed to other types of products. Some examples illustrating this chemistry are shown in Eq. 8.30,[5, 24, 25] 8.31,[25] 8.32,[26] 8.33[27] and 8.34.[28]

$$R-CH=C(SeR')(Me) \xrightarrow[Br_2,\ EtOH,\ \Delta]{H_3O^{\oplus}\ \text{or}\ Hg^{\oplus\oplus}} R-CH_2-C(=O)-Me \qquad (8.30)$$

$$C_8H_{17}CH=C(SeMe)(Me) \xrightarrow[(64\%)]{Br_2,\ 20^\circ C} C_8H_{17}CH=C(Br)(Me)$$

$$C_8H_{17}CH=C(SeMe)(Me) \xrightarrow[(51\%)]{n\text{-}Bu_3SnH,\ AIBN} C_8H_{17}CH=CHMe \qquad (8.31)$$

$$C_9H_{19}CH=CHSePh \xrightarrow[(95\%)]{HBr} C_9H_{19}CH_2CH(SePh)(Br) \qquad (8.32)$$

$$C_9H_{19}CH=CHSePh \xrightarrow[\text{or } PhSeO_2H\ (82\%)]{PhSe(O)OSe(O)Ph} C_9H_{19}CH(SePh)-CH(=O) \qquad (8.33)$$

$$CH_2=CHSePh \xrightarrow[(2)\ MeCOMe\ (72\%)]{(1)\ n\text{-}BuLi} C_5H_{11}CH(SePh)C(CH_3)_2OH$$

$$\xrightarrow[(81\%)]{[O]} C_4H_9CH=CHC(CH_3)_2OH \qquad (8.34)$$

Oxidation of vinyl selenides with one equivalent of *m*-chloroperbenzoic acid (mCPBA) gives the corresponding selenoxides, which can be used for the synthesis of acetylenes (Eq. 6.41) or cyclopropyl ketones (Eq. 8.35).[29]

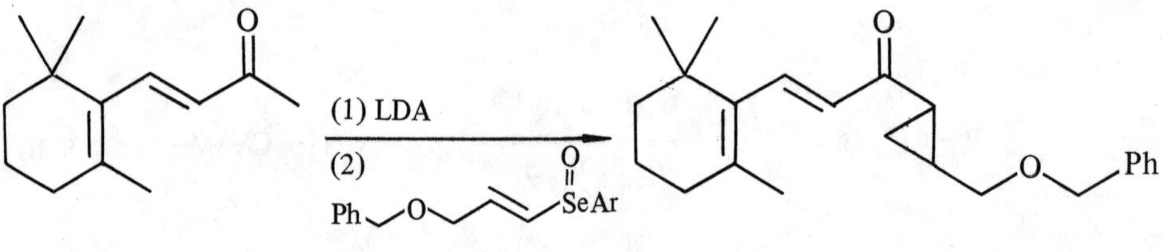

(8.35)

By using two equivalents of mCPBA vinyl selenides are converted to selenones, which have been used by Kuwajima for the synthesis of oxetanes (Eq. 8.36),[30] ethylenic ketones (Eq. 8.37);[31] acetylenic ketones (Eq. 8.38)[31] and substituted cyclopropanes (Eq. 8.38a).[117]

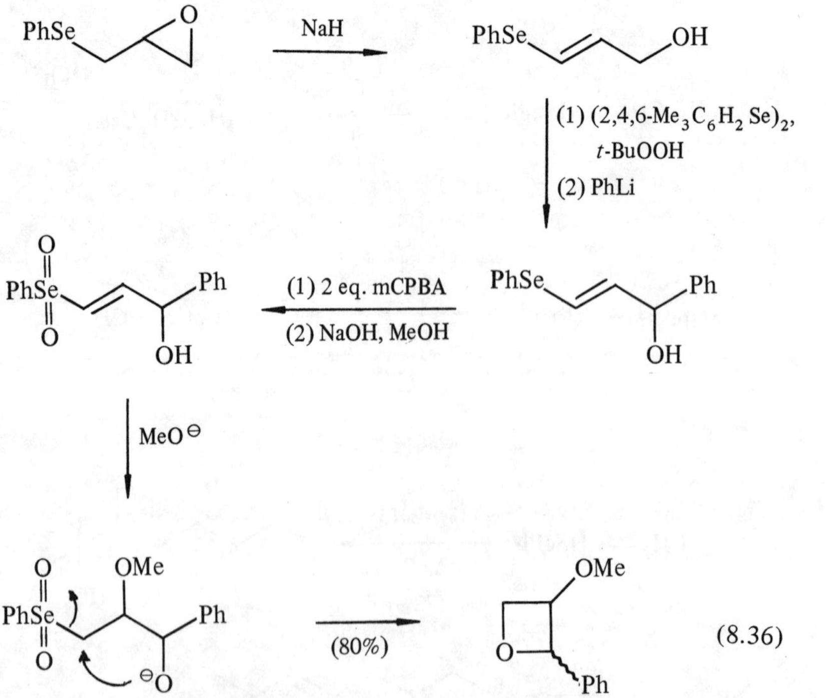

(8.36)

(1) n-BuLi (2) 2 eq. mCPBA; MeONa (86%)

(8.37)

NaH (84%)

(8.38)

$NCCH_2COOMe$, NaH (91%)

(8.38a)

8.3 Synthesis and reactions of silyl selenides

Phenyltrimethylsilyl selenide (**26**) can be prepared by reacting trimethylsilyl chloride with phenyl selenolate anion or with benzeneselenol in the presence of a rhodium catalyst, as shown in Eq. 8.39.[32–36]

$$PhSeM \xrightarrow{Me_3SiCl} \underset{\mathbf{26}}{PhSeSiMe_3} \xleftarrow[Rh(Ph_3P)_3Cl]{Me_3SiCl,} PhSeH$$

(M = Na, Li, Tl)

(8.39)

The presence of a hard acid component (Me_3Si) and a soft base component (PhSe) in this molecule gives it some unique properties, particularly towards oxygenated compounds.[37] Thus, it was found that carbonyl compounds can be converted efficiently to silylated phenylseleno derivatives by treatment with PhSe$SiMe_3$ in the presence of a Lewis acid ($MgBr_2$,[38] $ZnCl_2$,[32, 38] BF_3,[32] Ph_3P[32] Me_3SiOTf[39]). Some examples with aldehydes and ketones are given in Eq. 8.40, 8.41 and 8.42, taken from the work of Liotta *et al.*[32]

PhSeSiMe$_3$, ZnCl$_2$ (100%) (8.40)

PhSeSiMe$_3$, PPh$_3$ (100%) (8.41)

PhSeSiMe$_3$, PPh$_3$ (100%) (8.42)

Based on this type of reaction, Noyori developed a one-pot α-alkoxyalkylation of enones, illustrated in Eq. 8.43.[39]

PhSeSiMe$_3$, cat. Me$_3$SiOTf; CH(OEt)$_3$, cat. Me$_3$SiOTf; H_2O_2 (8.43)

The reagent **26** is in fact a carrier of the anionic $PhSe^{\ominus}$ group and it gives reactions typical of this species in the presence of fluoride anion ($F^{\ominus}$)[33] or a Lewis acid (ZnI_2,[37, 40] $TiCl_4$[41]). Examples are shown in Eq. 8.44,[33] 8.45,[33, 37] 8.46,[37] 8.47,[41] and 8.48.[40]

$PhSeSiMe_3$, KF, 18-C-6, THF (73%) (8.44)

$PhSeSiMe_3$, KF, 18-C-6, THF (81%) or $PhSeSiMe_3$, cat. ZnI_2 (79%) (8.45)

$PhSeSiMe_3$, cat. ZnI_2 (98%); $PhSeSiMe_3$, cat. ZnI_2 (77%) (8.46)

$PhSeSiMe_3$, cat. $TiCl_4$ (80%); (1) *n*-BuLi (2) $H_3O^{\oplus}$ (75%) (8.47)

$$\text{ClCH}_2\text{-epoxide} \xrightarrow[\text{cat. ZnI}_2\ (76\%)]{\text{PhSeSiMe}_3,} \text{ClCH}_2\text{CH(OSiMe}_3\text{)CH}_2\text{SePh} \quad (8.48)$$

Reaction of **26** with halogen (iodine, bromine or chlorine) gives the corresponding trimethylsilyl halide and diphenyldiselenide, which can be reacted *in situ* with various substrates. In the case of iodine the resulting Me_3SiI can act as a catalyst for the addition of **26** to a molecule, as exemplified in Eq. 8.49.[42]

$$\text{CH}_2\text{=CHC(O)CH}_3 \xrightarrow[(100\%)]{\text{PhSeSiMe}_3,\ \text{I}_2} \text{PhSeCH}_2\text{CH=C(OSiMe}_3\text{)CH}_3 \quad (8.49)$$

Detty has also prepared several other silyl selenides with a different trialkylsilyl group (Et_3Si, t-$BuMe_2Si$, t-$BuPh_2Si$). All of these compounds react with halogen to form the corresponding iodides, which react with epoxides and alcohols as indicated in Eq. 8.50 and 8.51.

$$\text{cyclohexene oxide} \xrightarrow[(92\%)]{t\text{-BuPh}_2\text{SiSePh, I}_2} \text{2-iodocyclohexyl-OSi}^t\text{BuPh}_2 \xrightarrow[(98\%)]{\text{DBU}} \text{cyclohex-2-enyl-OSi}^t\text{BuPh}_2 \quad (8.50)$$

$$\text{MeO-steroid-OH} \xrightarrow[(58\%)]{t\text{-BuMe}_2\text{SiSePh, I}_2\text{, pyridine}} \text{MeO-steroid-OSi}^t\text{BuMe}_2 \quad (8.51)$$

8.4 Miscellaneous applications of phenylseleno carbonyl compounds and their derivatives

Phenylselenoacetaldehyde (28) is a very useful synthon for homologation–functionalization of halides[43] and ketones.[44,45] It has been prepared from ethylvinyl ether (27) via an alkoxyselenation reaction, shown in Eq. 8.52.[43]

EtO ⟋ 27 —(PhSeBr, EtOH)→ (EtO)₂CH–CH₂SePh —($H_3O^{\oplus}$)→ OHC–CH₂SePh 28 (8.52)

Reaction of 28 with Grignard reagents (29) derived from primary, secondary or vinylic halides gives the homologated β-hydroxy selenides (30), which can be converted to olefins (31) (§4.6), epoxides (32) (§5.2) or phenylselenomethyl ketones (33) (§6.1), as indicated in Eq. 8.53. Examples of this methodology are given in Eq. 8.54[43] and 8.55.[43]

RMgX 29 + H(C=O)CH₂SePh 28 → RCH(OH)CH₂SePh 30; 30 —(§ 4.6)→ R–CH=CH₂ 31; 30 —(§ 5.2)→ R-epoxide 32; 30 —(§ 6.1)→ RC(=O)CH₂SePh 33 (8.53)

n-C₄H₉MgBr + H(C=O)CH₂SePh —(89%)→ n-C₄H₉CH(OH)CH₂SePh —(NCS, Me_2S, Et_3N (74%))→ n-C₄H₉C(=O)CH₂SePh (8.54)

(8.55)

Involvement of **28** in aldol-type reactions, followed by elimination of the resulting β-hydroxyselenides (**34**) to olefins was applied recently in a method for the preparation of α-vinyl substituted ketones (**35**, Eq. 8.56).[44,45] This procedure is useful for the synthesis of 1,5-diolefinic systems subjected to the Claisen–Cope or the oxy-Cope rearrangements, as illustrated in Eq. 8.57[44] and 8.58.[45]

(8.56)

(8.57)

Me$_3$SiO ... OHCCH$_2$SePh / ZnCl$_2$ (95%) ... MsCl, Et$_3$N (82%) ... (70%) MgBr ... KH, Δ (95%)

(8.58)

α-Phenylselenoketones (e.g. **36**) were converted to allyl selenides (**38**) by treatment with Me_3SiCH_2Li, followed by reaction of the resulting adduct (**37**) with stannous chloride (Eq. 8.59).[46]

36 —Me$_3$SiCH$_2$Li (51%)→ **37** —cat. SnCl$_2$ (82%)→ **38**

(8.59)

β-Phenylseleno esters (**41**) were obtained from allylic alcohols (**39**) and 3-phenylseleno orthopropionate (**40**) via a Claisen ortho ester rearrangement (Eq. 8.60).[47] Oxidation–syn elimination of **41** gives α,β-unsaturated esters, such as **42** (Eq. 8.61).[47] Alternatively, **40** was used in an approach to α-methylene lactones, such as **43** (Eq. 8.62)[47] by utilizing the phenylselenolactonization reaction (§7.2).

39 + **40** —Me$_3$CCOOH→ **41**

(8.60)

$$\xrightarrow{\mathbf{40}} \qquad \xrightarrow{H_2O_2} \mathbf{42} \qquad (8.61)$$

(8.61)

(8.62)

β-Phenylselenocarbonyl compounds, obtained by conjugate addition of phenylselenolate anion ($PhSe^{\ominus}$) to α,β-unsaturated carbonyl compounds (§4.1) are very useful synthetic intermediates.[114] They have been used in the synthesis of Δ^7-isoprostacyclin (**44**, Eq. 8.63)[48] and furanoprostacyclin (**45**, Eq. 8.64).[49]

O
O
R
OSi^tBuMe_2

(1) LDA
(2) PhSeCl
(3) H_2O_2

O
O
R
OSi^tBuMe_2

(1) PhSeNa
(2) DIBAL
(3) $BrMg(CH_2)_3CH{=}CH_2$

R =
OSi^tBuMe_2

HO
HO
SePh
R
OSi^tBuMe_2

(1) TsCl, pyridine
(2) 9-BBN
(3) H_2O_2

H
COOH
O
HO
OH

44

(1) $H_3O^{\oplus}$
(2) Pt, O_2

H
OH
O
R
OSi^tBuMe_2

(8.63)

(8.64)

Conjugate addition of arylselenols to cyclohexanones in the presence of cinchona alkaloids (cinchonidine, quinine) proceeds with asymmetric induction, giving optically active products, as shown in Eq. 8.65.[50]

(8.65)

8.5 Cleavage of carboxylic esters and methyl ethers

Uncomplexed sodium phenylselenolate, PhSeNa, prepared from PhSeSePh and sodium or from PhSeH and NaH, was found by Liotta[51] to cleave carboxylic esters (**46**) to the corresponding acids (**47**) via an S_N2 type reaction (Eq. 8.66). Examples are shown in Eq. 8.67[51] and 8.68.[51] Similar reactions with lactones are discussed in §4.4.

$$\underset{\mathbf{46}}{\mathrm{RCOOR^1}} \xrightarrow{\mathrm{PhSeNa}} \underset{\mathbf{47}}{\mathrm{RCOOH}} \qquad (8.66)$$

$$\mathrm{PhCOOCHMe_2} \xrightarrow[(99\%)]{\mathrm{PhSeNa}} \mathrm{PhCOOH} \qquad (8.67)$$

OMe ... COOMe $\xrightarrow[(100\%)]{\mathrm{PhSeNa}}$ OMe ... COOH (8.68)

Cava has shown that sodium benzylselenolate, $PhCH_2SeNa$, prepared *in situ* from $PhCH_2SeSeCH_2Ph$ and sodium borohydride, is a useful reagent for the regioselective demethylation of aryl methyl ethers (**48**) to the corresponding phenols (**49**, Eq. 8.69).

$$\underset{\mathbf{48}}{\mathrm{ArOMe}} \xrightarrow{\mathrm{PhCH_2SeSeCH_2Ph,\ NaBH_4}} \underset{\mathbf{49}}{\mathrm{ArOH}} \qquad (8.69)$$

Using this procedure, Cava was able to convert apomorphine dimethyl ether (**50**) to apocodeine (**51**, Eq. 8.70)[52] and selectively demethylate one methyl ether group of ocopodine (**52**) to form **53** (Eq. 8.71).[52]

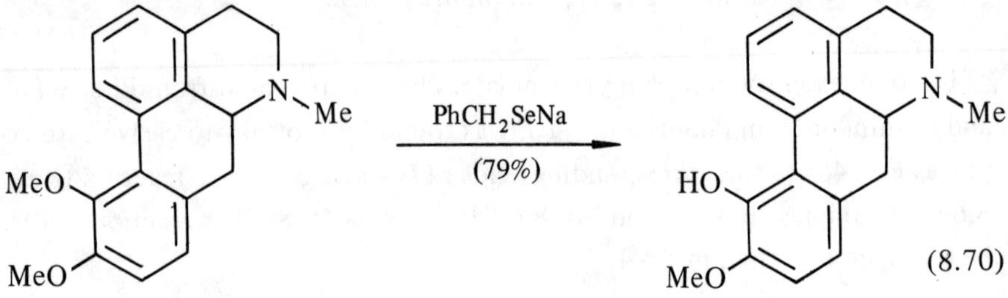

$$\xrightarrow[\text{(79\%)}]{PhCH_2SeNa} \qquad (8.70)$$

50 → **51**

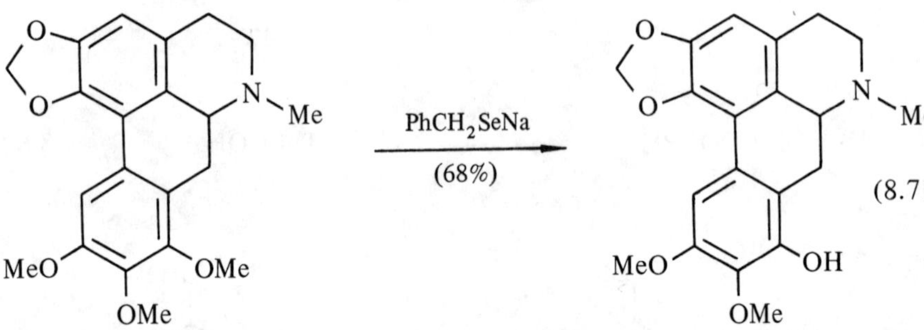

$$\xrightarrow[\text{(68\%)}]{PhCH_2SeNa} \qquad (8.71)$$

52 → **53**

Similar demethylations can also be done with MeSeNa[53] and PhSeNa.[54] An interesting example by Cava is shown in Eq. 8.72, where the methyl ether–epoxide **54** is converted directly to condensed furan **55**.

(8.72)

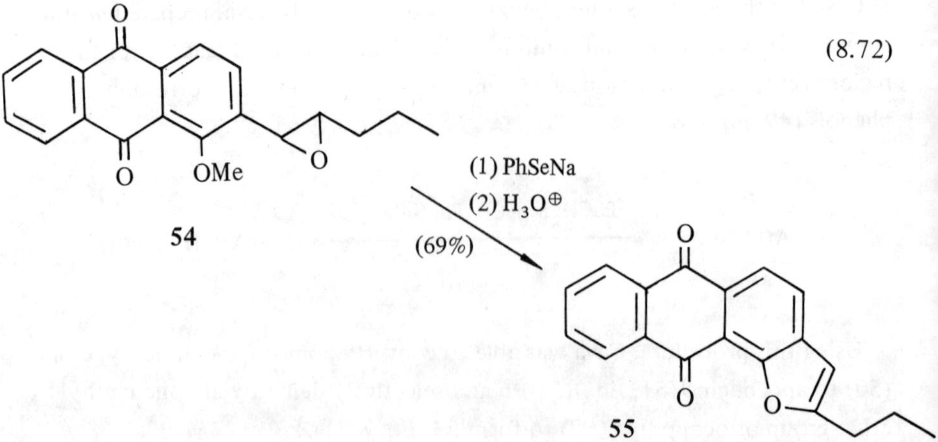

54 → (1) PhSeNa (2) $H_3O^{\oplus}$ (69%) → **55**

8.6. Reactions with sulfur compounds

The similar chemical reactivity between sulfur and selenium compounds allows the application of various selenium reagents for the oxidation or reduction of sulfur compounds. The determining factor for these reactions is the relative oxidation state between substrate and reagent. Some typical oxidative transformations are shown in Eq. 8.73,[55–58] 8.73a,[115] 8.74,[56–58] 8.75,[59] 8.76[60] and 8.77.[61]

$$\mathrm{RSR^1} \xrightarrow{\mathrm{ArSe(O)OOH}} \mathrm{RS(=O)R^1} \qquad (8.73)$$

$$\mathrm{PhCH_2SCH_2Ph} \xrightarrow[(89\%)]{(p\text{-}\mathrm{MeOC_6H_4})_2\mathrm{Se{=}O}} \mathrm{PhCH_2S(=O)CH_2Ph} \qquad (8.73a)$$

$$\mathrm{RSR^1} \xrightarrow{\mathrm{ArSe(O)OOH}} \mathrm{RS(=O)_2R^1} \qquad (8.74)$$

$$\mathrm{RSH} \xrightarrow{\mathrm{PhSeO_2H}} \mathrm{PhSeSR + RSSR} \qquad (8.75)$$

$$\mathrm{RSO_2NHNH_2} \xrightarrow{\mathrm{PhSeO_2H}} \mathrm{RSO_2SePh} \qquad (8.76)$$

$$\mathrm{ArSO_2H} \xrightarrow{\mathrm{PhSeO_2H}} \mathrm{ArSO_2SePh} \qquad (8.77)$$

The organoselenium-mediated oxidation of sulfides to sulfoxides (Eq. 8.73) and sulfones (Eq. 8.74) is done under very mild conditions and proceeds selectively even in the presence of double bonds. It has been used by Nicolaou in the synthesis of S–containing prostacyclins (Eq. 8.78),[57] and in his synthesis of ionophore antibiotic X–14547A (**56**, Eq. 8.79).[62a] The same transformation was also used by Ley in a similar synthesis of **56**.[62b]

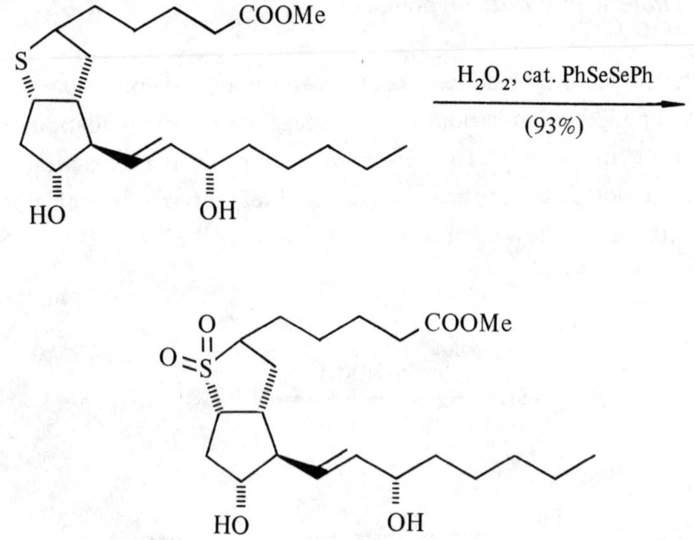

COOMe

S

HO OH

H_2O_2, cat. PhSeSePh

(93%)

(8.78)

COOMe

O

O=S

HO OH

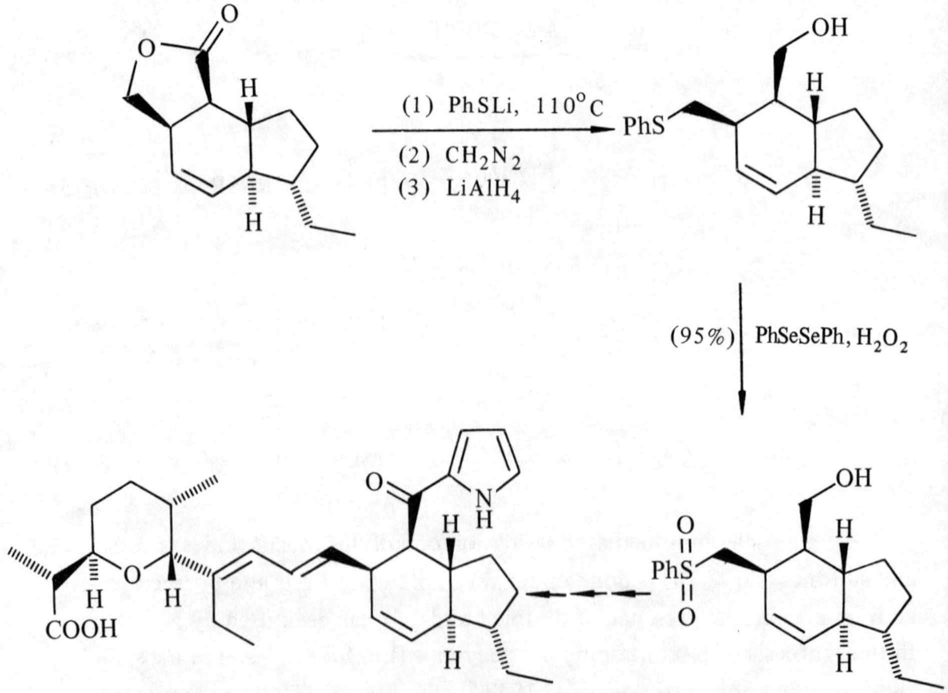

O

O

H

H

(1) PhSLi, 110°C

(2) CH_2N_2

(3) $LiAlH_4$

PhS

OH

H

H

(95%) PhSeSePh, H_2O_2

OH

H

O

PhS

O

H

O

N

H

H H

H H

COOH

H

56

(8.79)

Several organoselenium reagents were found to reduce sulfoxides to the corresponding sulfides, as summarized in Table 8.1. An example is shown in Eq. 8.80.[64]

Table 8.1. Methods of reduction of sulfoxides ($RS(O)R^1$) to sulfides (RSR^1).

Reagent	*Ref.*
$PhSeSePh{-}H_3PO_2$	63
PhSeH	64
$PhSeSiMe_3$	65
$Me_3SiSeSiMe_3$	66
$(EtO)_2P(O)SeH$	67
$B(SePh)_3$	68
$B(SeMe)_3$	68
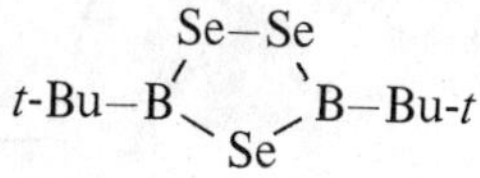 t-Bu–B(Se–Se)(Se)B–Bu-t (cyclic)	68

(8.80)

PhCH$_2$CONH, H, O, S, N, O, Me, COOCH$_2$CCl$_3$ → PhSeH, Δ (84%) → PhCH$_2$CONH, H, S, N, O, Me, COOCH$_2$CCl$_3$

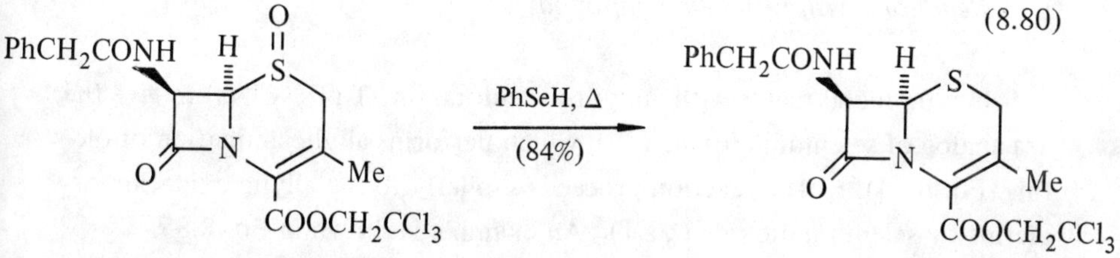

Benzeneseleninic anhydride [PhSe(O)OSe(O)Ph] was found by Barton to be an effective reagent for the removal of the thioacetal protective group. An illustrative example is shown in Eq. 8.81.[69]

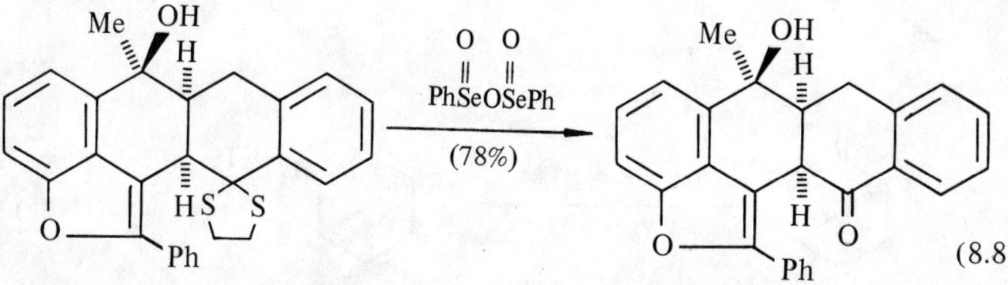

Me, OH, H, H, S, S, O, Ph → PhSe(O)OSe(O)Ph (78%) → Me, OH, H, H, O, O, Ph

(8.81)

Another reaction of benzeneseleninic anhydride with sulfur compounds is the conversion of thiocarbonyl compounds into their corresponding oxygen derivatives.[70] This type of transformation is also effected by dimethyl selenoxide [MeSe(O)Me].[71] Examples are given in Eq. 8.82[70] and 8.83.[71]

PhSeOSePh (71%)

(8.82)

MeSeMe (82%)

(8.83)

8.7 *Reactions with nitrogen compounds*

Selenium metal reacts with anhydrous chloramine-T (TsNClNa) to give the aza analog of selenium dioxide (57), which performs allylic amination of olefins (Eq. 8.84).[72] The reaction proceeds similarly to the allylic oxidation of olefins by selenium dioxide (§2.1). An example is shown in Eq. 8.85.[72]

TsN=Se=NTs **57** → NHTs

(8.84)

TsN=Se=NTs (82%) → NHTs

(8.85)

Treatment of 1,3-dienes with **57** gives the 1,2-diaminated products.[73] The reaction is believed to proceed via an initial [4+2] cycloaddition, as illustrated in Eq. 8.86.[73]

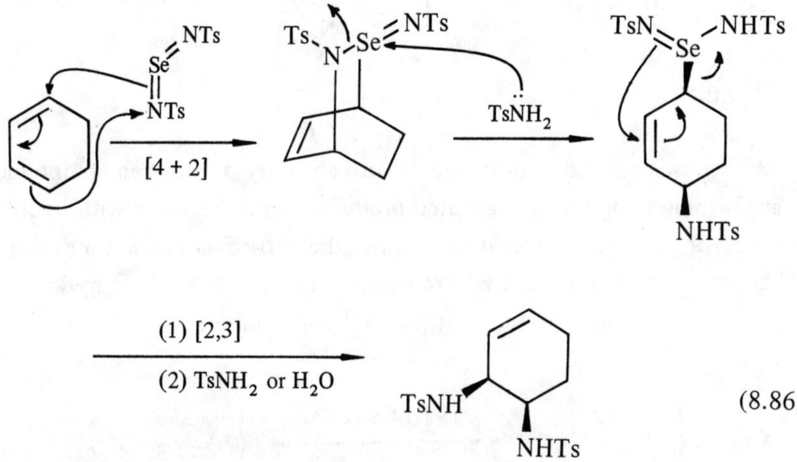

(8.86)

(37%)

Reaction of diphenyl diselenide (PhSeSePh) with chloramine-T (TsNClNa) forms a reagent that adds readily to olefins (**58**) giving, after reduction with sodium borohydride, β-sulfonamido selenides (**59**, Eq. 8.87).[74] Further elaboration of **59** gives selenium-free products, as shown in Eq. 8.88.[77]

(8.87)

58 **59**

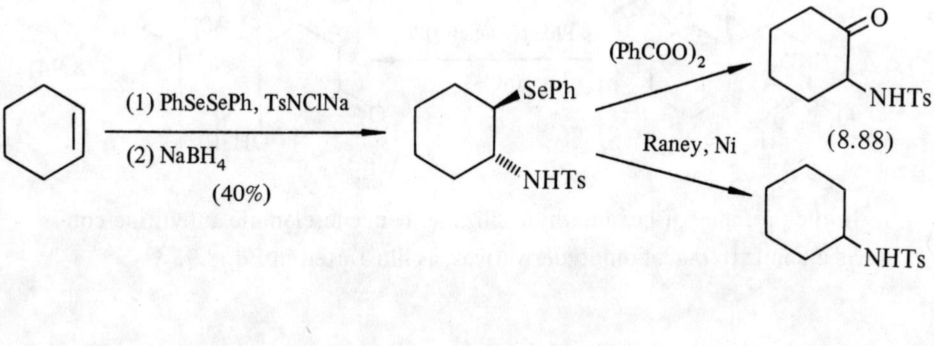

(8.88)

Selenium dioxide converts aldoximes (**60**) to nitriles (**61**, Eq. 8.89).[75]

$$\underset{\mathbf{60}}{RCH{=}N{-}OH} \xrightarrow{O{=}Se{=}O} \left[RCH{=}N{-}O{-}Se({=}O){-}OH \right] \longrightarrow \underset{\mathbf{61}}{RC{\equiv}N} \qquad (8.89)$$

Benzeneseleninic anhydride reacts with a variety of nitrogen compounds, transforming them into oxygenated products. Upon reaction with hydrazones, oximes or semicarbazones it regenerates the corresponding ketones (Eq. 8.90).[76] The same reagent oxidizes hydrazo compounds (Eq. 8.91),[76, 77] hydroxylamines (Eq. 8.92),[76] primary amines (Eq. 8.93)[78] and enamides (Eq. 8.94).[79]

$$R_2C{=}N{-}X \xrightarrow{PhSe(O)OSe(O)Ph} R_2C{=}O \qquad (8.90)$$

$X = NHR, OH, NHCONH_2$

$$RNHNHR \xrightarrow{PhSe(O)OSe(O)Ph} RN{=}NR \qquad (8.91)$$

$$RNHOH \xrightarrow{PhSe(O)OSe(O)Ph} RN{=}O \qquad (8.92)$$

$$RCH_2NH_2 \xrightarrow{PhSe(O)OSe(O)Ph} RC{\equiv}N \qquad (8.93)$$

$$\text{(steroidal enamide lactam)} \xrightarrow[(40\%)]{PhSe(O)OSe(O)Ph} \text{(N-H, OH, C=O hydroxy keto lactam)} \qquad (8.94)$$

In the presence of hexamethyldisilizane, benzeneseleninic anhydride converts phenols to *o*-acetamido derivatives, as illustrated in Eq. 8.95.[80]

$$\xrightarrow[\text{(64\%)}]{\text{PhSe(O)OSe(O)Ph, } (\text{Me}_3\text{Si})_2\text{NH}} \qquad \xrightarrow[\text{(83\%)}]{\text{Ac}_2\text{O, Zn}} \tag{8.95}$$

HO — O — NSePh — AcO — NHAc

Benzeneseleninic acid($PhSeO_2H$) or phenylselenenyl chloride (PhSeCl) catalyzes the rearrangement of *N*-arylhydroxamic acids to *p*-hydroxybenzanilides (Eq. 8.96).[81]

$$\xrightarrow[\text{or PhSeCl (70\%)}]{\text{PhSeO}_2\text{H}} \tag{8.96}$$

O — Ph — N — OH — O — Ph — NH — OH

Benzeneselenol (PhSeH) dealkylates tertiary, secondary and even primary amines at high reaction temperatures.[82] An example is presented in Eq. 8.97.[82]

$$\xrightarrow[\text{(97\%)}]{\text{PhSeH, 150° C}} \tag{8.97}$$

NHMe — NH_2

Reaction of phenylselenolate anion ($PhSe^{\ominus}$) with ditosylamides, obtained from primary amines, gives the corresponding selenides (Eq. 8.98).[83]

Ph–CH(CH$_3$)–NH$_2$ $\xrightarrow[\text{(2) NaH, TsCl}]{\text{(1) TsCl, Et}_3\text{N}}$ Ph–CH(CH$_3$)–N(Ts)$_2$ $\xrightarrow[\text{(70\%)}]{\text{PhSe}^{\ominus}\text{, }-80^\circ\text{ C}}$ Ph–CH(CH$_3$)–SePh

(8.98)

Phenylselenolate anion reacts with trisubstituted amines in the presence of ruthenium as catalyst to form the corresponding selenides in high yield.[84] Examples of this useful transformation are given in Eq. 8.99,[84] 8.100[84] and 8.101.[84]

MeO, MeO–C$_6$H$_3$–CH$_2$CH$_2$–N(SiMe$_3$)$_2$ $\xrightarrow[\text{(81\%)}]{\text{PhSe}^{\ominus}\text{, Ru}}$ MeO, MeO–C$_6$H$_3$–CH$_2$CH$_2$–SePh

(8.99)

NMe$_2$ $\xrightarrow[\text{(92\%)}]{\text{PhSe}^{\ominus}\text{, Ru}}$ SePh $\xrightarrow[\text{(90\%)}]{\text{H}_2\text{O}_2}$ OH

(8.100)

(8.101)

$\xrightarrow[\text{(2) H}_2\text{O}_2]{\text{(1) PhSe}^{\ominus}\text{, Ru, 100}^\circ\text{ C}}$ (67%)

OMe, OMe, N, O, O → NH, OMe, OMe, O, O

Several reactions of selenium compounds with diazomethane and ethyldiazoacetate have been observed and are given in Eq. 8.102,[85] 8.103,[86] 8.104[87] and 8.105.[88]

$$ArSO_2SePh \xrightarrow[(60\%)]{CH_2N_2,\, h\nu} ArSO_2CH_2CH_2SePh \quad (8.102)$$

$$RC(=O)SePh \xrightarrow{CH_2N_2,\, Cu} RC(=O)CH_2SePh \quad (8.103)$$

$$PhSeSePh + N_2CH \xrightarrow[(46\%)]{Cu} (PhSe)_2CH\text{—}COOEt \quad (8.104)$$

SePh, N₂CHCOOEt, CuSO₄ (35%), SePh, COOEt, COOEt, SePh

(8.105)

Phenylselenenyl chloride (PhSeCl) and diphenyl diselenide (PhSeSePh) react readily with α-diazocarbonyl compounds in the presence of boron trifluoride etherate to give disubstituted adducts.[89, 90] Treatment of these adducts with n-Bu_3SnH–AIBN affords the deselenylated products, as shown in Eq. 8.106.[89]

N⊖, N⊕, H, S, N, O, COOCH₂Ph, PhSeX, X = Cl, SePh, CH₂CH=CH₂, PhSe, H, X, S, N, O, COOCH₂Ph, n-Bu₃SnH, AIBN

$$\longrightarrow \quad \text{X-substituted }\beta\text{-lactam (H, H; S; N; O; } COOCH_2Ph) \qquad (8.106)$$

Carbonyl selenide (Se=C=O) was prepared from selenium metal and carbon monoxide (Eq. 8.107).[91] The reagent was found to react readily with amines to give the corresponding ureas and with hydroxylamines to give carbamates. Typical examples are shown in Eq. 8.108–8.111.[91]

$$\text{Se} \xrightarrow[\text{(2) } H_2SO_4]{\text{(1) CO, } Et_2NH} \text{Se=C=O} \qquad (8.107)$$

$$n\text{-BuNH}_2 \xrightarrow[(100\%)]{\text{Se=C=O}} n\text{-BuNHCONH}n\text{-Bu} \qquad (8.108)$$

$$Me_2NH \xrightarrow[(100\%)]{\text{Se=C=O}} Me_2NCONMe_2 \qquad (8.109)$$

$$H_2NCH_2CH_2NH_2 \xrightarrow[(86\%)]{\text{Se=C=O}} \text{imidazolidin-2-one (NH, NH, =O)} \qquad (8.110)$$

$$H_2NCH_2CH_2OH \xrightarrow[(70\%)]{\text{Se=C=O}} \text{oxazolidin-2-one (NH, O, =O)} \qquad (8.111)$$

8.8. Synthesis of halides

An organoselenium-mediated chlorination of olefins has been discovered by Sharpless.[92] The reaction involves the use of *N*-chlorosuccinimide (NCS) in the presence of PhSeCl, PhSeSePh, or *N*-phenylselenosuccinimide (N-PSS). Typical examples are shown in Eq. 8.112[92] and 8.113.[92]

NCS, cat. PhSeSePh (87%) (8.112)

SePh Cl N–PSS (87%) NCS (93%) (8.113)

Reaction of selenides with bromine gives the corresponding bromides with inversion of configuration.[93] Thus, the overall transformation of an alcohol to a selenide and then a bromide, proceeds with net retention of configuration (Eq. 8.114).[93]

OH PhSeCN, *n*-Bu_3P or $MeSO_2Cl$, PhSeNa SePh Br_2, Et_3N Br (8.114)

Phenylselenenyl chloride was recently found to be a regiospecific chlorinating agent of aromatic compounds, as shown in Eq. 8.115.[94]

HO Me PhSeCl (60%) HO Me Cl (8.115)

8.9 Deoxygenation of epoxides

Several selenium reagents can be applied to the direct deoxygenation of epoxides to olefins (Eq. 8.116). Alternatively, the epoxide can be converted to a hydroxyselenide, which then eliminates the elements of "PhSeOH" (§4.6). Table 8.2 shows the various direct methods of deoxygenation.

$$\text{epoxide} \longrightarrow \text{episelenide} \xrightarrow{-\text{Se}^\circ} \text{olefin} \qquad (8.116)$$

Table 8.2. Selenium reagents used in the deoxygenation of epoxides.

Reagent	*Ref.*
KSeCN	95
$Ph_3P{=}Se$	96
n-$Bu_3P{=}Se$	97
3,4-dimethyl-1-phenylphosphole 1-selenide (Me, Me; P; Ph, Se)	98
3-methylbenzothiazole-2-selone (S, =Se, N–Me)	99

An application of the reaction is found in Corey's synthesis of 12-HETE (**62**), as shown in Eq. 8.117.[100]

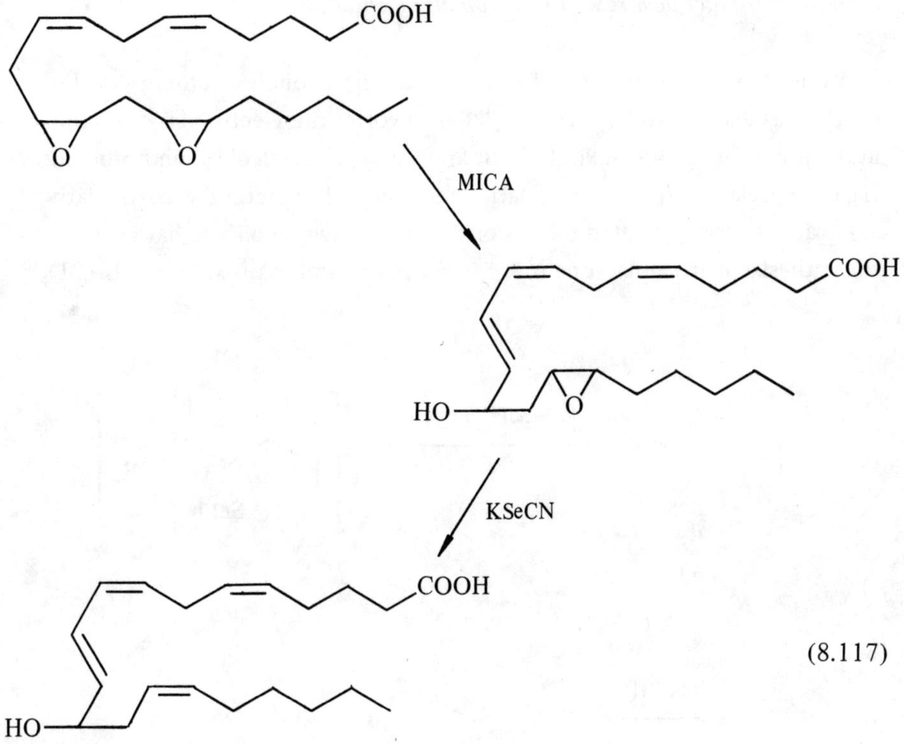

COOH

O O

MICA

COOH

HO O

KSeCN

COOH

(8.117)

HO

62

In an extension of this chemistry, Krief developed a method of stereoselective isomerization of disubstituted olefins (Eq. 8.118).[101]

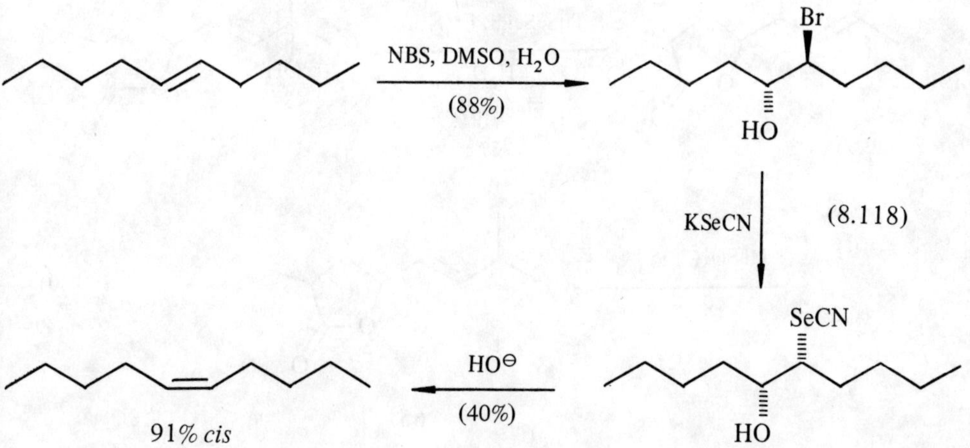

Br

NBS, DMSO, H_2O

(88%)

HO

KSeCN (8.118)

SeCN

$HO^{\ominus}$

(40%)

91% *cis* HO

8.10 Electrochemical reactions involving selenium

A one-step conversion of olefins (**63**) to allylic alcohols or ethers (**64**, Eq. 8.119) was discovered by Torii.[102–105] It involves the electrolysis of olefins in the presence of PhSeSePh, Et_4NBr and water (or an alcohol) and proceeds via an oxyselenation–deselenenylation sequence. The method is particularly applicable to trisubstituted olefins of the type shown in **63** and has been used to synthesize marmelolactone (**65**, Eq. 8.120)[102] and rose oxide (**66**, 8.121).[10]

PhSeSePh, Et_4NBr, ROH, Electrolysis

OR
SePh

63

(8.119)

–PhSeOH

OR

64

PhSePh, Et_4NBr, H_2O Electrolysis (85%)

O O OH O O

$MeSO_2Cl$, Et_3N, (83%)

(8.120)

O O

65

PhSeSePh, Et_4NBr,
MeOH,
Electrolysis,
(89%)

OH

OMe

OH

BF_3, OEt_2
(94%)

O

66

(8.121)

In a similar fashion, ketones were converted to their α-phenylseleno derivatives, presumably via their enolic forms, as indicated in Eq. 8.122.

O

OH

PhSeSePh, Et_4NBr, $MgBr_2$
H_2SO_4, MeOH,
Electrolysis,
(96%)

O

SePh

(8.122)

Other electrochemical reactions are given in Eq. 7.40 and 8.27.

8.11 *Synthesis of selenium analogs of natural products*

The presence of selenium in the structure of various compounds has been associated in some cases with biological activity. For this reason, many selenium-containing analogs of naturally occurring compounds have been prepared and studied. The most extensively selenylated group of natural products are probably the carbohydrates.[106] Some examples are shown in Eq. 8.123,[107] 8.124[108] and 8.125.[109]

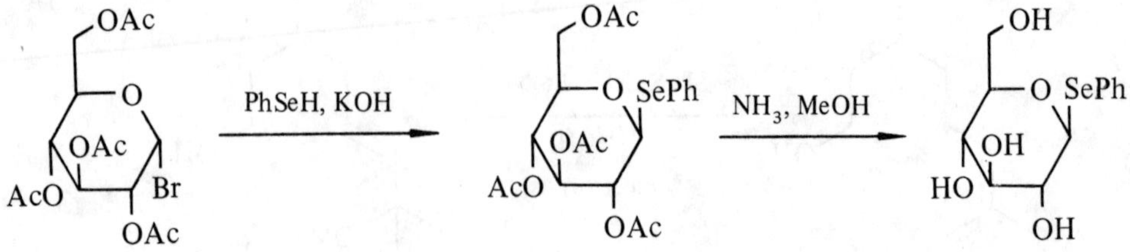

PhSeH, KOH; NH_3, MeOH

(8.123)

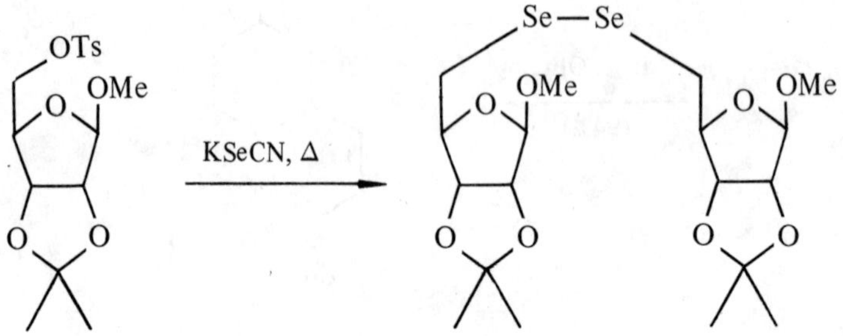

KSeCN, Δ

(8.124)

(1) $PhCH_2SeOH$, HCl
(2) TsCl, Py
(3) NaI, $BaCO_3$

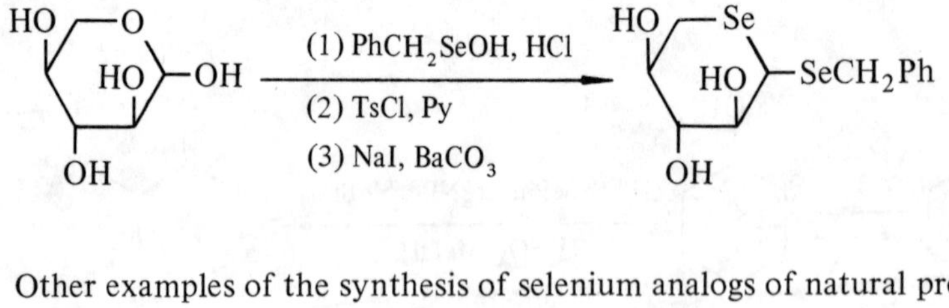

(8.125)

Other examples of the synthesis of selenium analogs of natural products are the syntheses of seleno-serotonin (**67**, Eq. 8.126),[110] and selenobiotin (**68**, Eq. 8.127).[111] Finally, the identification of selenomethionine in naturally occurring enzymes,[112] as well as the recognition of selenium as an important component of the enzyme glutathione peroxidase,[113] are worth mentioning.

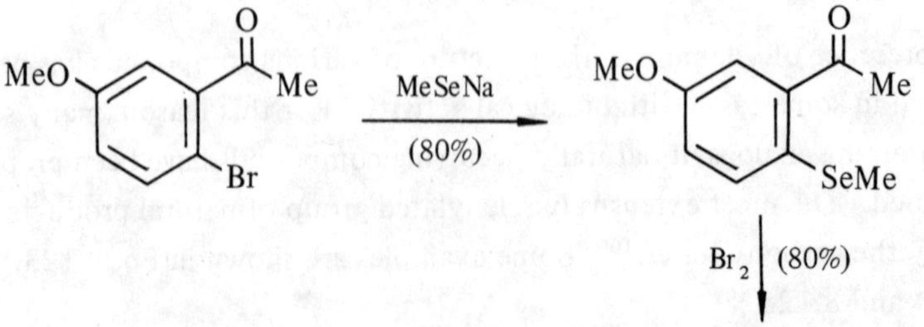

MeSeNa (80%)

Br_2 (80%)

(8.126)

Et$_3$N (98%)

(64%) Ph$_3$P=CHCN

(1) Py · HCl
(2) LiAlH$_4$, AlCl$_3$

67

MeOMgSeSeMgOMe

(8.127)

68

REFERENCES

1. Grieco, P. A., Yokoyama, Y., and Williams, E., *J. Org. Chem.*, **43**, 1285 (1978).
2. Grieco, P. A., Jaw, J. Y., Claremon, D. A., and Nicolaou, K. C., *J. Org. Chem.*, **46**, 1215 (1981).
3. (a) Back, T. G., Collins, S., and Kerr, R. G., *J. Org. Chem.*, **46**, 1564 (1981); (b) Back, T. G., and Collins, S., *Tetrahedron Lett.*, 2661 (1979).
4. Kozikowski, A. P., and Ames, A., *J. Am. Chem. Soc.*, **102**, 860 (1980); *J. Org. Chem.*, **43**, 2735 (1978).
5. Petragrani, N., Rodriguez, R., and Comasetto, J. V., *J. Organometal. Chem.*, **114**, 281 (1976).
6. Raucher, S., Hansen, M. R., and Colter, M. A., *J. Org. Chem.*, **43**, 4885 (1978).
7. Raucher, S., *J. Org. Chem.*, **42**, 2950 (1977).
8. Gröbel, B.-T., and Seebach, D., *Chem. Ber.*, **110**, 852, 867 (1977).
9. Reich, H. J., Willis, W. W., Jr., and Clark, P. D., *J. Org. Chem.*, **46**, 2775 (1981).
10. Dumont, W., Sevrin, M., and Krief, A., *Tetrahedron Lett.*, 183 (1978).
11. Sevrin, M., Dumont, W., and Krief, A., *Tetrahedron Lett.*, 3835 (1977).
12. Denis, J. N., and Krief, A., *Tetrahedron Lett.*, 23, 3407 (1982).
13. Reich, H. J., and Chow, F., *J. C. S. Chem. Comm.*, 790 (1975).
14. Comassetto, J. V., and Petragnani, N., *J. Organomet. Chem.*, **152**, 295 (1978).
15. Comassetto, J. V., and Brandt, C. A., *J. Chem. Res.*, 56 (1977).
16. Reich, H. J., Renga, J. M., and Trend, J. E., *Tetrahedron Lett.*, 2217 (1976).
17. Tomoda, S., Takeuchi, Y., and Nomura, Y., *Chem. Lett.*, 1715 (1981); *ibid.*, 253 (1982).
18. Bridges, A. J., and Fischer, J. W., *Tetrahedron Lett.*, **24**, 445, 447 (1983).
19. Miura, T., and Kobayashi, M., *J. C. S. Chem. Comm.*, 438 (1982).
20. Hooz, J., and Mortimer, R. D., *Can. J. Chem.*, **56**, 2786 (1976); *Tetrahedron Lett.*, 805 (1976).
21. Uneyama, K., Takano, K., and Torii, S., *Tetrahedron Lett.*, **23**, 1161 (1982).
22. Zima, G., and Liotta, D., *Synth. Comm.*, **9**, 697 (1979).
23. Shimizu, M., Ando, R., and Kuwajima, I., *J. Org. Chem.*, **46**, 5247 (1981).
24. (a) Hevesi, L., Piquard, J.-L., and Wautier, H., *J. Am. Chem. Soc.*, **103**, 870

(1981); (b) Piquard, J. L., and Hevesi, L., *Tetrahedron Lett.,* **21**, 1901 (1980).

25. Denis, J. N., and Krief, A., *Tetrahedron Lett.,* 23, 3411 (1982).
26. Dumont, W., Sevrin, M., and Krief, A., *Angew. Chem. Int. Ed. Engl.,* **16**, 541 (1977).
27. Cravadon, A., and Krief, A., *J. C. S. Chem. Comm.,* 951 (1980).
28. Raucher, S., and Koolpe, G. A., *J. Org. Chem.,* **43**, 4252 (1978).
29. Shimizu, M., and Kuwajima, I., *J. Org. Chem.,* **45**, 2921 (1980).
30. Shimizu, M., and Kuwajima, I., *J. Org. Chem.,* **45**, 4063 (1980).
31. Shimizu, M., Ando, R., and Kuwajima, I., *J. Org. Chem.,* **46**, 5246 (1981).
32. Liotta, D., Paty, P. B., Johnston, J., and Zima, G., *Tetrahedron Lett.,* 5091 (1978).
33. Detty, M. R., *Tetrahedron Lett.,* 5087 (1978).
34. Miyoshi, N., Ishii, H., Kondo, K., Murai, S., and Sonoda, N., *Synthesis,* 300 (1979).
35. Detty, M. R., and Wood, G. P., *J. Org. Chem.,* **45**, 80 (1980).
36. Detty, M. R., and Seidler, M. D., *J. Org. Chem.,* **46**, 1283 (1981).
37. Miyoshi, N., Ishii, H., Murai, S., and Sonoda, N., *Chem. Lett.,* 873 (1979).
38. Dumont, W., and Krief, A., *Angew. Chem. Int. Ed. Engl.,* **16**, 540 (1977).
39. Suzuki, M., Kawagishi, T., and Noyori, E., *Tetrahedron Lett.,* 22, 1809
40. Miyoshi, N., Kondo, K., Murai, S., and Sonoda, N., *Chem. Lett.,* 909 (1979).
41. Kuwajima, I., Hoshito, S., Tanaka, T., and Shimizu, M., *Tetrahedron Lett.,* **21**, 3209 (1980).
42. Detty, M. R., *Tetrahedron Lett.,* 4189 (1979).
43. Baudat, R., and Petrzilka, M., *Helv. Chim. Acta,* **62**, 1406 (1979).
44. Kowalski, C. J., and Dung, J.-S., *J. Am. Chem. Soc.,* **102**, 7950 (1980).
45. (a) Clive, D. L. J., Russell, C. G., and Suri, S. C., *J. Org. Chem.,* **47**, 1632 (1982); (b) Clive, D. L. J., and Russell, C. G., *J. C. S. Chem. Comm.,* 434 (1981).
46. Nishiyama, H., Hagaki, K., Osaka, N., and Hoh, K., *Tetrahedron Lett.,* **23**, 4103 (1982).
47. Raucher, S., Hwang, K.-J., and McDonald, J. E., *Tetrahedron Lett.,* 3057 (1979).
48. Shi, J. C., and Graber, D. R., *J. Org. Chem.,* **43**, 3798 (1978).
49. Nickolson, R. C., and Vorbrüggen, H., *Tetrahedron Lett.,* **24**, 47 (1983).
50. Pluim, H., and Wynberg, H., *Tetrahedron Lett.,* 1251 (1979).
51. (a) Liotta, D. N., Sunay, U., Santiesteban, H., and Markiewicz, W., *J. Org.*

Chem., **46**, 2605 (1981); (b) Liotta, D. N., Markiewicz, W., and Santiesteban, H., *Tetrahedron Lett.,* 4365 (1977).

52. Ahmad, R., Saá, J. M., and Cava, M. P., *J. Org. Chem.,* **42**, 1228 (1977).
53. Evers, M., and Christiaens, L., *Tetrahedron Lett.,* **24**, 377 (1983).
54. Ahmed, Z., and Cava, M. P., *J. Am. Chem. Soc.,* **105**, 682 (1983).
55. Faehl, L. G., and Kice, J. L., *J. Org. Chem.,* **44**, 2357 (1979).
56. Reich, H. J., Chow, F., and Peake, S. L., *Synthesis*, 299 (1978).
57. Nicolaou, K. C., Barnette, W. E., and Magolda, R. L., *J. Am. Chem. Soc.,* **103**, 3486 (1981); *ibid.,* **100**, 2567 (1978); *J. C. S. Chem. Comm.,* 375 (1978).
58. Nicolaou, K. C., Magolda, R. L., Sipio, W. J., Barnette, W. E., Lysenko, Z., and Joullié, M. M., *J. Am. Chem. Soc.,* **102**, 3784 (1980).
59. Kice, J. L., and Lee, T. W. S., *J. Am. Chem. Soc.,* **100**, 5094 (1978).
60. Back, T. G., and Collins, S., *Tetrahedron Lett.,* 2213 (1980).
61. Gancarz, R., A., and Kice, J. L., *Tetrahedron Lett.,* **21**, 1697 (1980).
62. (a) Nicolaou, K. C., Claremon, D. A., Papahatjis, D. P., and Magolda, R. L., *J. Am. Chem. Soc.,* **103**, 6969 (1981); (b) Edwards, M. P., Ley, S. V., Lister, S. G., and Palmer, B. D., *J. C. S. Chem. Comm.,* 630 (1983).
63. Günther, W. H. H., *J. Org. Chem.,* **31**, 1202 (1966).
64. Perkins, M. J., Smith, B. V., Terem, B., and Turner, E. S., *J. Chem. Res. (S),* 341 (1979).
65. Detty, M. R., *J. Org. Chem.,* **44**, 4528 (1979).
66. Detty, M. R., and Seidler, M. D., *J. Org. Chem.,* **47**, 1354 (1982).
67. Clive, D. L. J., Kiel, W. A., Menchen, S. M., and Wong, C. K., *J. C. S. Chem. Comm.,* 657 (1977).
68. Clive, D. L. J., Menchen, S. M., *J. C. S. Chem. Comm.,* 168 (1979).
69. (a) Cussans, N. J., Ley, S. V., and Barton, D. H. R., *J. C. S. Perkin I,* 1654 (1980); *J. C. S., Chem. Comm.,* 751 (1977); (b) Barton, D. H. R., Bielska, M. T., Cardoso, J. M., Cussans, N. J., and Ley, S. V., *J. C. S. Perkin I,* 1840 (1981).
70. Cussans, N. J., Ley, S. V., and Barton, D. H. R., *J. C. S. Perkin I,* 1650 (1980); *J. C. S. Chem. Comm.,* 393 (1978).
71. Mikolajczyk, M., and Luczak, J., *J. Org. Chem.,* **43**, 2132 (1978).
72. Sharpless, K. B., Hori, T., Truesdale, L. K., and Dietrich, C. O., *J. Am. Chem. Soc.,* **98**, 269 (1976); for the similarly reacting sulfur analog, see: Sharpless, K. B., and Hori, T., *J. Org. Chem.,* **41**, 176 (1976).
73. Sharpless, K. B., and Singer, S. P., *J. Org. Chem.,* **41**, 2504 (1976).

74. Barton, D. H. R., Britten-Kelly, M. R., and Ferreira, D., *J. C. S. Perkin I*, 1682 (1978).

75. (a) Sosnovsky, G., Krogh, J. H., and Umhoefer, S. G., *Synthesis*, 722 (1979); (b) Sosnovsky, G., and Krogh, J. A., *Synthesis*, 703 (1978).

76. Barton, D. H. R., Lester, D. J., and Ley, S. V., *J. C. S. Perkin I*, 1212 (1980); *J. C. S. Chem. Comm.*, 445 (1977); *ibid.*, 276 (1978).

77. Back, T. G., *J. C. S. Chem. Comm.*, 278 (1978).

78. Czarny, M. R., *J. C. S. Chem. Comm.*, 81 (1976); *Synth. Comm.*, **6**, 285 (1976).

79. (a) Back, T. G., Ibrahim, N., and McPhee, D. J., *J. Org. Chem.*, **47**, 3283 (1982); (b) Back, T. G., and Ibrahim, N., *Tetrahedron Lett.*, 4931 (1979).

80. Holker, J. S. E., O'Brien, E., and Park, B. K., *J. C. S. Perkin I*, 1915 (1982).

81. Frejd, T., and Sharpless, K. B., *Tetrahedron Lett.*, 2239 (1978).

82. Reich, H. J., and Cohen, M. L., *J. Org. Chem.*, **44**, 3148 (1979).

83. Müller, P., and Thi, M. P. N., *Helv. Chim. Acta,* **63**, 2168 (1980).

84. Murahashi, S.-I., and Yano, T., *J. Am. Chem. Soc.,* **102**, 2456 (1980).

85. Back, T. G., *J. Org. Chem.,* **46**, 5443 (1981).

86. Back, T. G., and Kerr, R. G., *Tetrahedron Lett.,* **23**, 3241 (1982).

87. Pellicciari, R., Curini, M., Ceccherelli, P., and Fringuelli, R., *J. C. S. Chem. Comm.*, 440 (1979).

88. Sharpless, K. B., Gordon, K. M., Lauer, R. F., Patrick, D. W., Singer, S. P., and Young, M. W., *Chem. Scr.,* **8A**, 9 (1975).

89. (a) Giddings, P. J., John, D. I., Thomas, E. J., and Williams, D. J., *J. C. S. Perkin I*, 2757 (1982); (b) Giddings, P. J., John, D. I., and Thomas, E. J., *Tetrahedron Lett.,* **21**, 395, 399 (1980).

90. Buckley, D. J., Kulkowit, S., and McKervey, A., *J. C. S. Chem. Comm.*, 506 (1980);

91. Kondo, K., Yokoyame, S., Miyoshi, N., Murai, S., and Sonoda, N., *Angew. Chem. Int. Ed. Engl.,* **18**, 691, 692 (1979).

92. Hori, T., and Sharpless, K. B., *J. Org. Chem.,* **44**, 4208, 4209 (1979).

93. Sevrin, M., and Krief, A., *J. C. S. Chem. Comm.,* 656 (1980).

94. Ayorinde, F. O., *Tetrahedron Lett.,* **24**, 2077 (1983).

95. Behan, J. M., Johnstone, R. A. W., and Wright, M. J., *J. C. S. Perkin I*, 1216 (1975).

96. Clive, D. L. J., and Denyer, C. V., *J. C. S. Chem. Comm.,* 253 (1973).

97. Chan, T. H., and Finkenbine, J. R., *Tetrahedron Lett.,* 2091 (1974).

98. Mathey, F., and Muller, G., *Compt. Rend.,* 281, 881 (1975).

99. Calo, V., Lopez, L., Mincuzzi, A., and Pesce, G., *Synthesis*, 200 (1976).
100. Corey, E. J., Marfat, A., Falck, J. R., and Albright, J. O., *J. Am. Chem. Soc.*, **102**, 1433 (1980).
101. Van Ende, D., and Krief, A., *Tetrahedron Lett.*, 2709 (1975).
102. Torii, S., Uneyama, K., and Ono, M., *Tetrahedron Lett.*, **21**, 2653 (1980).
103. Torii, S., Uneyama, K., and Ono, M., *Tetrahedron Lett.*, **21**, 2741 (1980).
104. Torii, S., Uneyama, K., Ono, M., and Bannon, T., *J. Am. Chem. Soc.*, **103**, 4606 (1981).
105. Torii, S., Uneyama, K., and Handa, K., *Tetrahedron Lett.*, **21**, 1863 (1980).
106. Witczak, Z. J., and Whistler, R. L., *Heterocycles*, **19**, 1719 (1982).
107. Bonner, W. A., and Robinson, A., *J. Am. Chem. Soc.*, **72**, 354 (1950).
108. Rabelo, J. J., and Es, T. V., *Carbohydrate Res.*, **30**, 381 (1973).
109. Blumberg, K., Fuccello, A., and Es, T. V., *Carbohydrate Res.*, **59**, 351 (1977).
110. Laitem, L., Thibaut, P., and Christiaens, L., *J. Heterocyclic Chem.*, **13**, 469 (1976).
111. Bory, S., and Marguet, A., *Tetrahedron Lett.*, 2033 (1976).
112. Esaki, N., Tanaka, H., Uemura, S., Suzuki, T., and Soda, K., *Biochem.*, **18**, 407 (1979).
113. Flohé, L., Günzler, W. A., and Loschen, G., in Kharasch, N., (ed.), *Trace Metals in Health and Disease*, Raven Press, New York (1979).
114. Miyashita, M., and Yoshikoshi, A., *Synthesis*, 664 (1980).
115. Ogura, F., Yamaguchi, H., Otsubo, T., and Tanaka, H., *Bull. Chem. Soc., Japan*, **55**, 641 (1982).
116. Sviridov, A. F., Ermolenko, M. S., Yashunsky, D. V., and Kochetkov, N. K., *Tetrahedron Lett.*, **24**, 4355, 4359 (1983).
117. Kuwajima, I., Ando, R., and Sugewara, T., *Tetrahedron Lett.*, **24**, 4429 (1983).

Index

α-Diazocarbonyl compounds, 279–280